COMMUNITY LIBRARY

3 2301 00072505 5

Ghost Ranch
Abiquiu, New Mexico
June 1980
For Mother - Martha Lee

D0573280

DATE DUE

FEB 27 1996		
OCT 7 1996		
AUG - 5 1997		
JUL 2 2 1998		
NOV 0 1 2001		
DEC 1 7 2001		
APR 0 2 2002		
GAYLORD		PRINTED IN U.S.A.

CAROLYN NIETHAMMER

WITH A FOREWORD BY Ann Woodin

Illustrations by Jenean Thomson

American Indian Food and Lore

COLLIER BOOKS

A Division of Macmillan Publishing Co., Inc.

NEW YORK

COLLIER MACMILLAN PUBLISHERS

LONDON

Copyright © 1974 by Carolyn Niethammer

All rights reserved. No part of this book may be reproduced or
transmitted in any form or by any means, electronic or mechanical,
including photocopying, recording or by any information storage
and retrieval system, without permission in writing from
the Publisher.

Macmillan Publishing Co., Inc.
866 Third Avenue, New York, N.Y. 10022
Collier Macmillan Canada, Ltd.

Library of Congress Cataloging in Publication Data

Niethammer, Carolyn.
  American Indian food and lore.

  Bibliography: p.
  1. Indians of North America—Southwest, New—
Food. 2. Ethnobotany—Southwest, New. I. Title.
[E78.S7N53 1974 b]   581.6′1′09701   74-8304
ISBN 0-02-010000-7

Third Printing 1978

American Indian Food and Lore is also published in a
hardcover edition by Macmillan Publishing Co., Inc.

Design by Bob Antler
Printed in the United States of America

To those ancient Indian women
who searched the arid lands for food
and labored over its preparation
—and to those who remember.

Foreword

One of the most perplexing questions the North American Desert poses the newcomer is: How did all those Indians once said to have roamed its stony expanses, ever find enough to eat? This was a primary consideration during my first years of camping in the middle of a remote lava wilderness that, to my eye, couldn't keep a fly alive—much less the group of Indians who used to live there. In the dead of night I sometimes would wake up and, after admiring the current overhead constellation, muse on the possibility that in the morning our vehicle would not start; and there we would be. How long could we stay alive, living off the land, I wondered? Not long, I concluded. Not long at all. Then those supposedly 'primitive' Indians took on formidable proportions.

First they had to be sufficiently aware and attentive to their surroundings in order to accumulate the necessary knowledge of the plants and animals with whom they shared this exacting parcel of land. Secondly, that knowledge had to be precise, detailed, and applicable to unusual situations or changes in conditions. Little if any margin for ignorance or error here. Nor was there anyone to blame but yourself should you fail to find dinner. A very heavy trip, to quote a local expression.

So when Carri Niethammer's manuscript came into my hands holding within it the chance of an answer to that much puzzled-over question, I was delighted. I distinctly remember the late summer afternoon I first met Carri beside a clump of prickly pear. In one hand she held a pair of forceps with which she was breaking off the ripe red fruit. In the other hand she held a bucket. A few weeks later Carri came by my house and offered me some dried bits of fruit from that same prickly pear. A neighbor's child was with me and (somewhat gingerly) we each sampled a piece. It was delicious.

Now I have read the aftermath of her labors in field and kitchen. Once published I will place *American Indian Food and Lore* among those books to which I keep wanting to refer; and I will have it with me on my next desert camping trip. Twenty pages of footnotes and a six-page bibliography reinforced my initial observation that this work was thoroughly researched and accurate in details. It is also informative about more than plants native to this area and how to cook them. Of particular interest to me were the descriptions of the resourceful and indomitable Indian ladies who, before the days of the corner grocer, managed to keep themselves and their families fed in what I have sometimes heard called an inhospitable wasteland. Which is waterless to boot. "Why, even a crow has to pack a canteen when he flies over there."

Obviously of primary concern in implementing a book of recipes based on wild plants, is the matter of how to identify the desired plants when out in the field. For this purpose Carri has included a description of each plant used. These descriptions have been supplemented and embellished by the line drawings of Jenean Thomson which add an important visual dimension. In fact, this business of identification is crucial enough that, at the risk of repeating what Carri may already have said, I would like to contribute a few words of my own.

The wise collector learns the common poisonous plants to be found in his favorite hunting grounds along with those for which he is looking. If one day you are very hungry and in doubt, test a likely-looking plant by carefully tasting (not swallowing) a small piece of it—first raw and then cooked. And then await 'developments.' This latter is a harmless appearing word which, as used here, may mean anything from numbing your mouth to burning it, as the experience Carri so vividly describes amply confirms. Another procedure is to persuade a knowledgeable friend to accompany you on your initial forays—particularly advisable when it comes to what I call 'weeds' but the afficionado refers to as 'edible wild greens.' Some of these have the sneaky knack of appearing quite different during various stages of their life cycles, so that what you may consider to be an old friend when flowering, may look like another plant altogether when young and tender.

Joseph Wood Krutch patiently performed this service for me every spring. After the fourth year of identifying the same plants, he suggested I start a plant-press to, as he aptly phrased it, 'impress' on my mind the plants about which I was enthusiastic. I have followed his advice and now have a useful (and pretty) plant-book of my own making. Still, the safest advice in these matters is, as Carri put it: "Don't eat it unless you know what it is."

I would like to append an additional fact to the section on the agave. When this plant sends up its flowering stalk, it has come to the end of its life. To gather it, therefore, is not the act of depredation it might first appear to be. New young plants are usually well on their way, having sprung from the root-system of the dying parent.

Several times I accompanied Carri on her collecting trips. For me these were always instructive. I remember the day we dug up an agave and dragged it down a rocky hillside, and the day we gathered the flaming clusters of ocotillo blossoms. So I especially enjoyed and appreciated those places where Carri incorporates specific personal experience into her instructions, such as how to successfully pull up the roots of a cat-tail. Or how to remove the thorns from cholla buds and prickly pear pads. Or how to gather the fruit from the saguaro. This is all very practical and valuable information. I do admit to being timid of the recipe which begins "Select small barrel cactus. Remove spines and outer layers with long sharp knife."

Eating for survival and eating for pleasure are two very different things. What may be tasty to a hungry Papago may be unpalatable to me. Or how a Papago woman may have to spend the hours of her days is not the same as how I may chose to spend mine. Though lately, reading of food shortages and watching the climbing prices in the local market, I may well have to change my 'chose' to her 'have to.' Anyway, I have tried grinding mesquite beans in a metate and my short-lived efforts dramatically impressed me with the tediousness of their tasks, as well as with the skill necessary to execute them.

Which brings me to reiterate the point that this book is of value beyond the recipes it contains. It is full of ethnic information related to the plants used in the recipes that Carri has gleaned and assembled in a readable form. Yet I do think it instructive for every desert dweller to prepare and eat at least one meal from the ingredients he himself has gathered. It is another method of learning more about the land where you live. Sometimes a subtle method—such as the information your fingers pick up in handling various textures, or your tongue picks up as it tastes the sweetness of a prickly pear 'tuna' and the clinging pungency of mesquite beans.

Furthermore, such pursuits are often garnished by unexpected incidents. When about to pluck a barrel cactus fruit, you may find a cheeky ground squirrel in the process of beating you to it. I once encountered a skunk over an item we both fancied and, after a brief discussion as to whose larder was whose, I retreated. I also know where a honey tree is. But I recall in minute detail the attempts Pooh Bear made to help himself to the bees' hoard. His ingenious maneuver, as you will recall, of floating through the air underneath a blue balloon in hopes that the bees would mistake him for a friendly black cloud hanging harmlessly in the sky, had uncomfortable results.

When I leave the supermarket with a carload of groceries and drive through a desert that Carri's book assures me abounds with edibles, I sometimes feel like the tourist visiting a foreign land who eschews native foods in favor of stuff brought from home. A Pakistani friend told of an American trailer-caravan who passed by so provisioned that not one morsel of a native dish need be consumed.

Carri's account of how the Yuman women once spent their days leads me to reflect again on an old persistent and enigmatic theme (a theme underlying the Women's Liberation Movement too). Namely: How do I come to know myself as a woman? Two pertinent pieces of advice spring to mind: Don't ask, rather—look to yourself, for "you" are what is available; and, Begin with the obvious. Suddenly every daily activity becomes a rich source of information, from the way I dress to the way I touch my child, from the way I feel a man's touch to the way I prepare and serve food.

In her capacity of food-gatherer and preparer, the Indian woman's sense of worth, contrary to mine, was reinforced daily in doing her woman's things, for everyone knew full well that without her the family might not eat. Moreover these activities put her in contact with primordial feelings that are as powerful, insistent, and molding to her as to me. Inexpressable knowings I sense deep in my bones when I place a platter of food before another—to do with sustenance, with vulnerability (mine in the preparation, the other's in the partaking), with relationship (beware of the woman whose man you feed).

Carri states that a Hopi girl proposes marriage to the boy by leaving a plate of 'piki' bread on his doorstep. This is not only a provocative example of the proverb "A way to a man's heart is through his stomach," but further evidence of the often disguised fact that a woman does the choosing. Whether I admit it or not, food assumes a vital place in my relationship to a man, and in this I am no different than the Indian woman. To us both it is how we most palpably nourish. As I grow older I increasingly trust such body knowledge and when I mock or deny it, I find myself uprooted and confused, in fact, 'non-nourishing' to those around me.

A few years ago I was kneeling on the ground beside a fire stirring a pot of soup. We were camped in the Arabian desert and, having felt myself for days slipping on the surface of unfamiliar sights, that simple atavistic rhythm of stirring a pot returned me to myself, and I felt grounded.

I venture to guess that my predilection for camping recognizes and yearns for what the Yuman woman knew. By reducing, however briefly, the back-home clutter to a few pots and two meals a day, the shapes-of-things (myself included) stand out more clearly. I have noticed that in the wilderness the necessary tasks become spontaneously and by choice somewhat traditionally divided—my sons haul the water and chop the firewood, and I do the cooking. I think it no accident that in a time characterized by increasing personal alienation where a technology has separated us from tasks that help us to describe ourselves, there are those who have turned to the land and a simpler mode of life. The impulse is profound and significant however debatable the method chosen to answer it may be.

Living off the land when the land is a desert is indeed an arduous and time-consuming business, as Carri's book shows. When my head is spinning out disjointed fragments like a machine run amok, I am perverse enough to envy the ancient Indian woman's very busyness. I would like to think that her outwardly-ordered days left no time or energy for poking around in the tangled corners of one's head.

And I envy them something else: the conviviality of their daily tasks. I remember the afternoon spent in the woman's section of a Bedouin tent that stretched taut over my head like a bird wing. The owner's three wives were with me. Here, three stones on the sand formed the kitchen stove and the only visible implements were two long-spouted coffee pots, a large blackened kettle, and one bowl, blue on the outside. I watched the intertwining movements of the women as they prepared the next meal, and felt a harmony, a tranquility, and . . . an ineffable hum of companionship deeper than anything I myself had yet experienced. Thinking of the few heavily shrouded women I had seen in the market-place who walked along without so much as a peephole to see out of, I was reminded of the word 'sisterhood,' and what it might mean here, in this far-away lonely land which had been grating me on its edges all those weeks of our traveling. Sitting in the company of those dark-eyed women with tattooed faces, while our respective sons and husbands drank coffee on the other side of the partition, I was strangely comforted and assuaged.

*　　*　　*

A few months ago some unexpected guests arrived just before lunch. My inside larder being almost bare, I hurried down a nearby wash to where, on a moist bank under a live oak, I had seen growing a lush patch of Miner's lettuce, also called Indian lettuce and *Montia perfoliata*. I soon had a large bowl-full.

"My," exclaimed the guest after her second helping, "this is about the most delicious salad I have ever eaten." For this moment I am grateful to Carri's book which has turned my eyes newly to the desert outside my window and extended the dimensions of its landscape.

—Ann Woodin

Acknowledgments

The real spirit of this book comes from the Indian women who patiently recounted recipes and demonstrated techniques for making traditional Indian dishes. Without their generous help this volume could not have been written. Their names and tribal affiliations are Loretta Blatchford, Ella Tsinnie, and Katherine Shirley, *Navajo*; Nancy Howard and Bessie Kewenvoyouma, *Hopi*; Pauline Goodmorning, *Taos Pueblo*; Minnie Marshall, *Havasupai*; Lula Bowanie, *Zuni*; Molly Manuel, *Papago*; Grace Mitchell, *Yavapai*; and Daisy Johnson, *Apache*. Roy and Martha Ward were helpful in introducing me to many of these women.

Library research was made easier by the assistance of the staff at the Arizona State Museum Library, especially Daphne Scott.

Thanks also go to Dr. Mike Robinson, who identified plants and checked the manuscript for botanical accuracy; to Dr. Thomas Hinton, who cheerfully explained many matters of anthropological significance; to Julie Szekely for reading the manuscript and for eliminating scores of unnecessary words and restructuring many awkward sentences; to Lloyd Findley, who helped with the field work, took photographs for the artist, and offered encouragement; and to my parents, who had faith that I would finally finish it.

Contents

Most readers will not have previously eaten the plants which are the subject of recipes in this book. We urge you to be absolutely certain that you have properly identified the plants and that you follow the recipes and warnings in the book when preparing them. If you have an allergic history, you should be guided accordingly. We cannot make any claims for the effectiveness of any of the medicinal uses to which the plants described in the book have been used and suggest that you consult a physician before using them medicinally.

Introduction

Now another day is coming,
Awake from slumber,
Look toward the east,
See the rising of the sun,
Which means another day to toil.
Another day to hunt for meat,
To put the seed in the ground
That the yield might be good,
So our people may not go hungry.
The great Father provided us the sun
To give life to our earth so
That it might give us a good yield.
And that we might see
To hunt our game for meat.
So arise and make use of the day

And do not get in the way
Of the women as they go about
Fixing up the camp and the
Needed task of preparing meals for you.
Many moons, many suns have come and gone
Since our forefathers here on this same ground
Toiled and struggled so that we might
Enjoy life today.
So let us not waste this day.
But get your tools, go out to the field, or
Take down your bow and arrows
And go after the game, so that
Your family will not be in need of meat.
So now I hope you will strive
To make this day the best in your life.[1]

In Pima villages in what is now central Arizona, early in the morning of each day, just before sun-up, a man with a clear voice could be heard a mile away intoning a prayer something like the one above. The man spoke about five minutes and when he finished talking the village inhabitants began to move about at their tasks. The Pimas were farmers who lived near the Gila and Salt rivers and who relied on an extensive irrigation system to grow their crops. Although they did cultivate some of their food, wild plants were still widely utilized. Aboriginal dwellers of the arid lands of the Southwest were reverent toward the plants and animals that nourished them. They had come close to facing hunger too often to take the presence of plentiful food for granted.

The groups that lacked enough water for agriculture or had not developed agriculture as a basis for their livelihood had to be especially attuned to the seasons. These people had to know where each variety of edible plant grew and when its fruits became ripe. An error of a week or so might mean the family or band would go without a particular food, a supply of seeds or plants for basketry, or a needed tool for an entire year, until the required plant would once again come into its particularly useful stage.

Take berries, for example. If they are not harvested soon after they become ripe, hungry birds, also seeking to survive on parched lands, will take their share—which might be every berry on the bush.

A University of Michigan archeologist, Kent V. Flannery, has aptly written:

> We no longer think of the preceramic plant-collectors as a ragged and scruffy band of nomads; instead they appear as a practiced and ingenious team of lay botanists who know how to wring the most out of a superficially bleak environment. Nor do we still picture the Formative peoples as a happy group of little brown farmers dancing around their

Aboriginal Territories of Southwestern Indians

cornfields and thatched huts; we see them, rather, as a very complex series of competitive ethnic groups with internal social ranking and great preoccupation with status, iconography, water control and the accumulation of luxury goods.[2]

Although many of the plants within a particular group's range were extensively utilized they were not exploited. Man was not simply extracting energy from his environment, but participating in it, and his use of each genus was part of a system that allowed that genus to survive, and sometimes flourish, in spite or because of heavy utilization.[3]

An example of this is the harvesting and eating of cactus fruits. No matter how intensive the gathering it does not diminish the available stands of cactus. Unless specially processed, the seeds of the prickly pear and saguaro almost inevitably survive the human digestive tract and escape in the feces, thus allowing the plant maximum seed dispersal.

Of course, the Indians' relationship to the plant world is a product of centuries of accumulated knowledge. The time span between the instance a single individual, driven by hunger or curiosity, tried a new plant and when that plant was fully adopted into the culture might be hundreds of years. Sometimes a few people died or became seriously ill after eating a plant raw, and only many years later did someone else discover that same plant to be harmless, perhaps even delicious, when cooked.

Thus, the longer a particular group has inhabited an area, the more fully they know and utilize the local flora. The Hopis, for instance, use for food, medicine, drugs, tools, or the like almost every wild plant that grows on their mesas. The Navajos, in contrast, use some wild plants but not all

XX

that grow in the area in which they live. To the ethnologist this difference, combined with other evidence, shows that the Hopis have occupied their mesas for many centuries, while the Navajos are relative newcomers.[4]

The Zunis have fully adopted plants into their world view. These pueblo people live with their plants—considering them a part of themselves. According to Matilda Cox Stevenson, an anthropologist who did work with the Zunis in the early 1900's, the initiated traditionally could talk with their plants and the plants, in turn, with them. Plants were considered sacred by the Zunis, for they believed some of them had been dropped to the earth by the Star people; some were human beings before they became plants; others were the property of the gods. Vegetation is symbolized by blue-green on the sacred dance kilts worn by the personators of the rainmakers and there are many other designs on fabrics, ceramics, and ceremonial objects symbolizing the fullness of the earth.[5]

About the Apaches and their relationship to the flora of their region Keith H. Basso writes:

> Although the Western Apache engaged in subsistence farming, their economy was based primarily on the exploitation of a wide variety of natural resources by hunting and gathering. It is estimated that agricultural products made up only 25 per cent of all the food consumed in a year, the remaining 75 per cent being a combination of meat and undomesticated plants. Because they could not rely on crops throughout the year, the Western Apache did not establish permanent residences in any one place. In fact, except for early spring, when there was planting to be done, and early fall, the time of harvest, they were almost constantly on the move.
>
> In May the people deserted low-altitude winter camps in the Salt and Gila River valleys and treked overland to farm sites in the White Mountains. Here they seeded small plots (generally about a half-acre in size) of corn, beans and squash. Once this was finished, large gathering parties set out in search of saguaro fruit, the prickly pear, and the Apache staple, mescal (agave). Older people unwilling to make the trek, the disabled, and a few children remained behind to cultivate the fields.
>
> Acorn and mesquite beans were collected in July and August and by September the fruit of the spanish bayonet (*Yucca baccata*) was ready for picking. In the early fall the Apache returned to their farm sites, harvested their crops, and spent most of October and November gathering pinyon nuts and juniper berries. As the food was brought in, it was either eaten on the spot or stored in baskets for the months ahead. Hunting was best in the late fall, and it was not until a good supply of game had been secured that they moved again towards their winter camp, thus completing the annual cycle.[6]

The southwestern Indians' relationship to their foods and their methods of gathering it was also intricately involved with their relationships with one another. Arid lands produce gatherers. Desert plants produce great quantities of seeds, and thus manage to propagate themselves in an area of uncertain growing conditions. These abundant seeds provide an easily gathered harvest for the Indians. Seeds were doubly attractive because they could be easily ground and made into bread or mush. Because collecting

An Apache woman
spreads baked and
mashed mescal pulp
on a bed of grass.
After the pulp has
been dried in the
sun, it can be stored
for future use.
(*Arizona State Museum
—University of Arizona,
Goodwin Collection.*)

seeds in the dry lands must be done quickly and intensively for a maximum harvest, it leads to group activity. When many families converge on an area to harvest a wild crop, some organization is needed to regulate and coordinate the actions of the group. Positions of authority emerge and are delegated to particular individuals.[7]

In many cases the work of gathering and preparing what was gathered fell to the women and children, while the men were out hunting for game. Picking squawberries, gathering prickly pears, cleaning the stickers off cholla buds, and the constant grinding of grains and seeds at the metate, or grinding stone, were all rather tedious work. How much more quickly the time went when the women worked together and had someone with whom to chat!

I experienced this need for companionship when I was gathering ingredients to test the recipes in this book. Standing in the hot sun, picking sticky, fuzzy squawberries all alone was fun for the first five minutes, but the succeeding half-hour that it might take to fill my basket was boring and trying. That is, unless there was a friend or acquaintance who wanted to learn about wild foods who happened to be picking with me—which happened only rarely. But when there was a friend, the time flew as we chatted our way from bush to bush.

While the Indian women worked together their gossiping not only relieved their boredom, but it intensified their culture and helped keep the village together. There were no newspapers in those small villages to tell the inhabitants of the latest events, yet it was extremely important for every individual to know what was happening. Information of importance to the women was undoubtedly passed around at the daily work parties. Mores and

xxii

morals also came under close scrutiny at these gatherings. A woman would think twice about deviating from the accepted code of behavior if she knew her conduct would be the big item for gossip at the next day's grinding party.

I was flooded with the feeling for what these women's gatherings might have been like, while camping out at the pueblo ruin of Hawikuh, several miles from the present town of Zuni. We had arrived in Zuni too late in the afternoon for me to try to find a woman willing to talk about Zuni cooking that day, so we drove out to the extensive ruin to spend the night. Hawikuh, around 1300 to 1680, was a large pueblo inhabited by about six hundred people, according to archeologists who did some digging there from 1917 to 1923. Today nothing remains there but a heap of rubble and the faint outline of some houses. However, at the edge of the ruin I did find one real gem: a large rock, as big as a kitchen table, with grinding troughs all around it. Looking at the rock, I began to have a vivid picture of the Hawikuh women of ancient times all sitting around it, pounding away at whatever they were preparing and talking up a storm. I could see the dark stones of the pueblo looming behind and the green fields stretching in front of them. I wanted that rock, but it weighed several tons, and anyway it is against the law to remove such artifacts. Nevertheless, I have claimed it in my heart.

The next day, in Zuni, luckily, and totally unexpectedly, I witnessed a modern-day version of the same type of interaction between women. After asking around the town to find out who was a good cook and who might tell me about Zuni cooking, I was sent to the home of Lula Bowanni. I am always reluctant to go to the house of a total stranger asking for information, and usually try to arrange an introduction. But in this situation, there was just no way this could be done. So I just went up to the modern and attractive house and knocked on the door, and was greeted with amazing friendliness and cordiality. And, by happy accident, Mrs. Bowanni and her relatives were in the midst of preparing a feast.

The "old grandfather" of the family had died, and so, according to the beliefs of the tribe, on the fourth day of his death he had to be fed. Several of the granddaughters and nieces were in the living room kneading huge pans of bread dough which would be formed into loaves to be baked in the adobe beehive oven in the backyard. (Mrs. Bowanni's recipe for the bread began: "three 25-pound bags of flour . . . ")

Then I was invited into the kitchen to taste the mutton stew in preparation. It was a very modern kitchen, with a refrigerator, an upright freezer, a washer-dryer combination, and a blender. But on top of the washer was a pan full of blood and entrails waiting to be made into blood sausage, and in the corner of the kitchen neatly laid on clean newspapers were the head and legs of the recently butchered sheep. The old and the new combined beautifully into one another.

When I returned later in the afternoon after collecting some plants, all the women were sitting around the dining table having a late lunch. Everyone from grandmother to little tots was there. Nothing like the isolation of the suburban housewife here. Those women were having a great time, eating and laughing. And, of course, carrying an ancient culture into modern times.

However, except for feasts similar to the one which Mrs. Bowanni and her relatives were preparing, the arid-land Indians of ancient times did not have complicated recipes or serve elaborate meals. Usually they just had one food at each meal—a mush made of the dominant staple. Whatever the day's gathering brought in, whether greens or seeds or berries, was added to the mush for variation.

Total adaptation to the environment required great ingenuity in food preparation. Even without having any utensils that would seem suitable for cooking to us, these ingenious first Americans could cook a variety of foods. Several arrowweed stalks tied together made excellent stirring implements for mush and loosely woven baskets did very well for sifters. Preceramic groups prepared many foods by boiling and stewing in their baskets. Their baskets were so closely woven as to be able to hold liquids. Hot stones were dropped into the liquid and replaced by others as they cooled until the necessary temperature was achieved. Hide receptacles also were used in the same way.

In the Mogollon culture of 600 to 800 A.D. cooking was done in large pits filled with rocks and heated with fire. While the rocks were heating, food was wrapped in some insulating material, such as grass. The rocks were removed from the hot pit, and the wrapped food was placed in the pit and covered with the hot rocks, which were topped with earth.[8]

A woman could cook a dish with several different wild ingredients only if all the ingredients were in season at the same time or if she had managed to dry and store a sufficient quantity of each food to last until the next season, when new plants could be combined for the dish. Some foods were very difficult to store.

The Hopi interestingly rationalized what was inevitable in their diet and included the eating of common and rare foods right into their code of manners. They classified foods into two general categories: staples or widely available foods, and other foods which were more scarce.

The scarce or more unusual foods, which were usually tastier than the monotonous staples of corn and beans, often served the function of flavoring and adding variety to everyday foods. Included in this classification were wild and domestic meat, wild plants, salt, sugar, chili, and onions. It was considered bad manners to eat large quantities of these foods even when they were in good supply, and children were taught this rule early in life. The Hopi custom thus ensured that their small supply of such foods would be enough to satisfy their entire population.[9]

Let us take a typical day in the life of a woman of one of the Yuman tribes along the Colorado River, say a member of the Mohave or Maricopa groups. Domestic duties began before sunrise. The woman rose early and immediately began her preparations for the morning meal. She then went to fetch a day's supply of water from the river, carrying a large pot balanced on a cloth ring on her head. Fetching water was women's work and the women always made a point of having the storage jars filled before dawn.

The preparation of meals was laborious. The first meal was served at sunrise as soon as the rest of the family had risen. Any woman seen preparing breakfast after dawn was considered lazy. After breakfast the women often left to go into the desert to gather foodstuffs. Children remained at home to tend the fire and build up a store of ashes to be used in cooking.

When the women returned home in the middle of the afternoon they began preparing the evening meal. Much time had to be spent pounding mesquite beans in the mortar or grinding corn at the metate (stone-grinding trough). Wild seeds also were ground. Because the ability to grind quickly and well was the most important accomplishment of a housewife every prospective bride was set to work grinding by her future mother-in-law to prove her proficiency.

Foods were half-broiled, roasted, or boiled, and then dried for storage. To conserve water, always a consideration on the desert, foods that could be roasted were prepared by being baked in ashes rather than boiled. For baking, the ground was heated and the food was placed among the ashes with a few live coals raked back over the article. Boiling was, however, the mainstay of cooking. Mesquite, corn, seeds, pumpkins, squash, and beans usually were cooked in the ceramic pots.

Cooking was done with mesquite firewood. The men went out to break up the wood but the women had to carry it home.[10]

Food was shared among several families. One woman would prepare a large amount of some vegetable and then portion it out to all the other women preparing meals. They, in turn, would bring her some of whatever they were cooking. The evening meal was eaten as soon as the food was ready, usually before sunset.

Fastidious eaters washed their mouths out before and after the meal. Whoever cooked served the others first. If the men had prepared small game they had killed they served their wives first. However, since most meals were of vegetables and had been prepared by the women, the men ordinarily ate first.

To the east of the Yumans lived the Pima farmers, who relied on river water to irrigate their crops. About every fifth year the Gila River failed in midwinter. Without water in the river and canals there could be no crops, so the Pimas had to resort to eating only wild foods. They gathered saguaro and mesquite and ranged far for the plants that would produce fruits or seeds even in dry seasons. When these, too, failed the Pimas had to make long journeys to the mountains in search of animal foods, roots, berries, and edible agaves. These trips were very dangerous because it took the Pimas into the territory of their enemies, the Apaches. At other times the very abundance of water was a disaster. Floods destroyed the canals and swept away growing or ripening crops.[11]

Similar to the Pimas were their southern neighbors, the Papagos. The Papagos, not living near an important river, had much less agriculture. The ordinary Papago diet consisted of dried foods: cereals, dehydrated vegetables, and sun-dried meat. Fresh foods were available for only short periods, then they were consumed in quantity.

Castetter and Underhill in *The Ethnobiology of the Papago Indians*[12] tell us that the year-round staple for the Papago was flour made from corn or various wild seeds and beans. The technique with all except the sticky mesquite was to parch, sun-dry, and store the whole seeds in sealed jars. The parching was done at the time of gathering and was a part of the storage technique to prevent mildew. The seeds were not ground into flour until just before they were used.

Flour was most often served in the form of a gruel. There were two kinds: waka, the Mexican pinole, which served warriors and people on journeys with both food and drink, and ator (atole) which was the gruel served at home. Waka was made simply by mixing parched corn with water. For atole the flour was boiled with water and salt. The pot in which the gruel was cooked usually included a bit of something extra. If there was too little left for a meal, more flour of any kind was added and the whole cooked. Any number of combinations were possible, the favorite being corn-meal with saguaro seed flour, which was slightly oily and sweet.

Castetter and Underhill did their research in the early 1930's when the Indians could still remember how things were when more traditional food customs still pervaded. Gruel or bread, they were told, was the basis of every meal and the usual addition was boiled dehydrated vegetables. All greens, like agave hearts and cholla buds, had been pit-baked before storing; root crops and the cultivated squash and beans had been simply sun-dried. All of these were soaked in water before cooking to freshen them and remove dirt, then boiled with water and salt.

Fresh greens also were boiled, but again, anything that was juicy enough was roasted. In a time of extreme water scarcity, dehydrated vegetables could also be baked in ashes after the first soaking.

The sweets, highly prized, were dried fruits, fruit jam, fruit syrup, and honey. Saguaro and prickly pear fruits could be dried and stored in jars, but not for long, because they became wormy. During the summer, honey was plentiful.

Navajo customs were somewhat different, as investigated by Flora L. Bailey in the 1940's.[13] Hauling water and cutting firewood were men's duties, undertaken by women only when necessary. The women of the household generally worked together to prepare meals, but anyone present could have been drafted to help, even the men. Some men would offer help, watching food lest it burn, when the women were busy elsewhere. Children, both boys and girls, were expected to help by tending the fire, running errands, and washing dishes. Flora Bailey was told by the men she talked to that they had learned to cook by watching and thought it quite customary for men to cook.

It was very improper for a woman to stand while cooking. It was also thought impolite for either a man or woman to stand in the hogan. However, a woman would sometimes stand by an outdoor fire to tend food if other things needed attention at the same time. She would bend from the waist, knees straight, holding her full skirts back from the flames with one hand.

Firewood was usually juniper; if it was not available pinyon was used. Dry juniper wood makes a good fire with a pleasant odor and the clear smoke from the juniper does not blacken the cooking pots. Pinyon makes a pitchy smoke that discolors the cooking vessels.

Among the Indians food was often presented as a gift. Among the New Mexican Tewas* all cooked foods were appropriate presents from women to men or between the women of households. The proper present from a man to a woman was game, firewood, or clothing.

In the *Ethnobotany of the Tewa Indians*[14] we learn that on the eve of a festival the women sent new bread and pies made of dried peaches and melons to their neighbors, the present folded in a cloth and carried under the bearer's headshawl. The receiver of the gift emptied the dish, wiped it, and gave it back to the bearer; or, more ceremoniously, washed dish and cloth before returning them the next day.

At festivals women and girls carried bread, cakes, boiled meat, chili con carne, and coffee to the kiva or as refreshment for the dancers. On the Day of Kings (January 6) when the dancers performed before the houses of the newly appointed officials of the pueblo, the officials' wives bestowed boiled meat, bread, and boiled pumpkin on the watching audience. At some dances women and girls brought baskets full of meal and set them down before their favorite dancers, who were supposed to give a present of game in return. One men's dance at Santa Clara had fallen into disuse "because the men are afraid to dance; there are some women capable of giving a man nine baskets of meal, and now that rabbits are so scarce, he would be ruined in buying meat to pay them."[15]

Food figures heavily in Hopi courtship and marriage customs. If a girl was interested in a boy, she took a Maiden's Cake of blue corn meal and removed enough of the boiled meal to make two small narrow packages about the size of your middle finger. These two small cakes were done up in fresh white husks, tied in several places, and then tied together, side by side. The little cakes were made at Bean Dance time in February, when the people are in the seed kivas (sacred underground rooms). As the young boys filed past the maidens during the dance, the girl would slip a pair of these little packets into the boy's hand.[16]

It is also up to the Hopi girl to propose to the Hopi boy. When the girl decides whom she wants to marry she prepares a large plate of blue corn wafer bread called piki and, accompanied by her mother and maternal uncle, she goes to the home of the young man and leaves the piki on the doorstep. If the youth and his family take the wafer bread in, the proposal of marriage is accepted. But if they leave it on the doorstep, a member of the girl's family fetches the unaccepted plate of bread so the young woman will not be embarrassed.

Although Indian foods are healthful and natural, many of the dishes do not taste like typical "American" food, which is amply spiced and has

---

* Who are linguistically related groups of Pueblo Indians living along the Rio Grande and in northern Arizona.

distinct flavors. But Indian foods are edible, and many of the dishes are very good. As with any foreign cuisine, some will prefer certain dishes more than others, probably because they more closely resemble the kinds they are used to.

However, some Indian recipes that I tried while researching this book I have judged truly inedible. Perhaps I did not have the correct information on how to prepare them (the knowledge may be lost) or perhaps my palate is too attuned to modern foods. The recipes for these dishes have been omitted here. I would not encourage anyone to spend a laborious day in the desert collecting little seeds that are bad-tasting. Some of these foods might have been more attractive to someone who was starving, a familiar situation to the Indians. In very lean times when old people ate only once in three days in order to leave food for the children, no digestible substance was likely to be overlooked.

Several periods of severe famine occurred among the Tewas between 1840 and 1860. The people were so hungry that if they had a good piece of rawhide—such as would be used now for shoe soles—they would roast it, grind it, and make it into bread.[17]

While a present-day American might balk at the high calorie count of a saguaro fruit, to a Papago homemaker who had the problem of gathering enough food to keep her family alive it was indeed a blessing to be able to collect such a rich food.

None of the recipes here contain anything but natural, wholesome ingredients and the dishes described in the following chapters are usually considered tasty. The recipes usually make small or moderate quantities, so you can sample the dish before cooking up a large portion. All the recipes can be doubled, if you wish, except for the jams and jellies which give better results when made in small quantities. Most of the recipes here are authentically Indian; some I have changed a little to adapt to modern tastes. Also included are some recipes that are completely modern, yet make use of plants that grow wild in arid regions. With a little experimentation you will be able to adapt many of these foods to your personal tastes.

Although it is fun to have a real Indian grinding stone or metate (mine is one of my favorite possessions), you can bend to modern technology, like many of the Indians I visited, and use a hand food mill or an electric blender. I use a Corona food mill which was recommended by our local health food store.

The first reaction of many people when they think of eating foods from the wilds is the fear of being poisoned. I have eaten all the foods in this book but have never been poisoned or made ill. But I was guided by evidence, either from anthropological literature or from Indians themselves, that these plants were traditionally used as foods. As a word of warning, however, I must tell you of the one unpleasant experience I did have.

After testing and nibbling on unusual plants for two years, I had become a little overconfident. One day, while moving to another house, I inadvertently moved a large, jungle-type of house plant called diffenbachia. The plant had grown awkward and top-heavy and the small trunk broke off near the base. My first reaction was acute regret that I had killed such a lush, beautiful plant, but then, curious, I looked at the broken stem. Inside the thin green outer skin the inner flesh was succulent-looking and white, much like sugar cane. Surely this must be good to eat. I just touched it to my tongue for an instant—and ran quickly to the kitchen sink to rinse it out. My entire mouth was immediately aflame! Nothing could stop it. I had never been poisoned before but I immediately knew it had happened then. With my thick, pained, by-now paralyzed tongue I managed to blurt out instructions to get help. I was rushed to the local university hospital. Fortunately, the person who took me to the hospital, a biologist himself, had the presence of mind to take along the plant for identification.

Of course, I was the first case of its kind the resident and nurse had seen. The plant was identified and the poisoning attributed to oxalic acid crystals. The only treatment was to take antihistamine and wash out the mouth with cold milk. The experience had the effect of keeping me quiet for a day and a half. The acute pain subsided in two days, the general numbness in two more days—but it was a week and a half before the numbness left my lips and the tip of my tongue—the only parts that had actually touched the plant.

The moral of this story is obvious—don't eat it unless you know what it is! Each plant in this book has an accompanying illustration, and should not be too hard to identify. I don't think that any of them is likely to hurt the average person if used as I have suggested.

Other books that are helpful in identifying plants that can be eaten are *Wild Edible Plants of the Western United States* by Donald Kirk, *Common Edible and Useful Plants of the West* by Muriel Sweet, and *Edible Native Plants of the Rocky Mountains* by H. D. Harrington. *Arizona Flora* by Thomas H. Kearney, Robert H. Peebles, and collaborators is complete and precise (readers of this volume will find a previous knowledge of botanical terms necessary).

Each plant included in *American Indian Food and Lore* is listed in alphabetical order under its most-often-used common name. Other common names and the scientific name for each plant are also given. It is important to know the scientific name of a plant because a plant often has many common names and sometimes the same common name will be used for two or more very different plants. But each plant generally has only one Latin two-part scientific name by which it is known all over the world. The first word, which is capitalized, names the genus, or group; the second word, not capitalized, names the species or kind, a subdivision of the genus. An example is *Agave palmerii*, or abbreviated, *A. palmerii*. *Agave* is the genus, *palmerii* the species. This is similar to the names people use. We say Sam Brown or Sally Brown. The genus name corresponds to Brown and the species name tells us which of all the Browns we are talking about. Thus, plants with the same genus name are closely related.

In some cases, I have listed more than one scientific name for a plant. In these instances the species are so similar that for our purposes they are interchangeable.

But knowing every common name and the scientific name of a plant does not necessarily mean that you can take the plant home for dinner. Most states have laws against the removal of some of their native plants from their natural habitat; the collection of seeds and fruits are not regulated.

For the purposes of recipes in this book, we are concerned only with regulations pertaining to agaves, barrel cactus, yucca, and wild onion. These plants are on the protected plants lists of both Arizona and New Mexico, which means that the entire organism cannot be collected unless specific regulations are observed. In Arizona, for example, a permit and a tag must be obtained for each plant you wish to collect. If the plant is on private land, you must also have the permission of the landowner. In New Mexico, you must have permission from the owner or agency supervising the land and no protected plants may be collected within 400 yards of any public highway. Violation of the law is considered a misdemeanor in both states and fines of up to $300 may be levied for each offense. So check the laws in your state before digging up any wild plants.

The land was the Indians' supermarket—supplying all their needs: groceries, medicines, eating utensils, clothing, tools, home-building materials, and so on. Many of the plants were used in a number of ways. After the recipes for each plant, I have listed other uses of that plant, including medicinal uses. Although modern pharmacologists have discovered that many of these primitive remedies did contain active medicinal elements, no research has been made in this area and I offer the notes on medicinal uses only for curiosity.

I

Cactus and Cactuslike Plants

agave

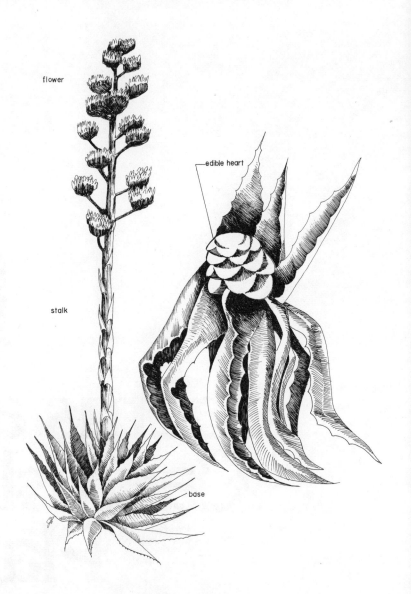

flower

edible heart

stalk

base

ALSO CALLED: Mescal and century plant.
Agave palmerii and *A. parryi*.

HABITAT and DESCRIPTION: The pre-flowering agave could be described as looking like an artichoke with an asparagus spear coming out of the center. The center stalk grows rapidly, sometimes as much as a foot a day, and eventually sends out lateral branches at the tip, each branch terminating in a large cluster of golden flowers. Flowering occurs in the late spring and early summer. The edible species are found in dry ground in Arizona, New Mexico, southeastern California, southern Utah, and northern Mexico, at elevations of 3,000 to 8,000 feet. The plant flowers once, then dies. Some plants live a long time before flowering—thirty or forty years—hence the name "century plant."

The century plant or agave was of enormous importance to the Indians of the Southwest; it provided food, fiber, and medicine. Its sweet taste made it a great favorite among many of the groups that could obtain it.

Among tribes using the agave in various ways were the Havasupais, Hopis, Yavapais, Maricopas, Papagos, Walapais, Kaibabs, and the White Mountain, Chiricahua, and Mescalero Apaches. The Mescalero considered it a staple—thus the name given to them by the Spaniards, who called the agave *mescal*. The Pimas used the agave mainly in times of famine. They would have liked to use it more often but they feared encountering their enemies, the Apaches—a possibility when they ventured even the shortest distance from home.

A great deal of ceremony accompanied the Apache mescal plant-gathering and -baking excursion. The first day of an expedition was usually spent gathering and preparing oak-digging sticks, getting a store of firewood, and digging the deep pits in which the mescal was roasted.

2

The plants were dug out by sticks about three feet long with one end flattened. These were pounded with a rock into the stem of the plant just below the base. The plant could then be removed easily. A broad stone knife was used to chop off the leaves. The naked crowns were bulbous, white in color, and from one to two feet in circumference.[1] If any of the thin-leaved varieties of agave were found by the gathering party they were rejected; it was believed that these varieties had little nutritional value and blistered the mouth.

The pits in which the crowns were baked were 10 to 12 feet in diameter, 3 or 4 feet deep, and lined with large flat rocks. (Hikers still find these pits while walking in remote areas.) One to 3 dozen crowns were roasted at once, depending on the size of the pit and how many women were working together.

According to Apache custom a cross was made with black ashes on the largest rock, which was placed in the center of the pit. Other rocks were then piled on the flat stones and on this rock mound oak and juniper wood was placed. The fire was lit before dawn, and usually died down by noon. Traditionally the women who were cooking did not light the fire; this task is always done by a small boy.

When the fire died down, moist grass was laid on the hot stones. The largest mescal crown was selected, crossed with cattail pollen, and thrown into the pit, followed by the rest of the crowns. Then the youngest child present stood at the edge of the pit and threw four stones in.

After being covered with beargrass and a thick layer of earth, the crowns were left to bake. During the 1½- to 2-day period that the mescal was roasting the Indians refrained from drinking (so as to prevent rain) and women were not to engage in any sexual activity with their husbands. If the crowns were not fully baked the first time they were checked, the condition was blamed on the incontinence of some of the women.[2]

Apparently the outings were considered fun, as they afforded an opportunity to get away from the village and more hum-drum chores. Daisy Johnson, a White Mountain Apache about seventy, has supplied a wealth of information on old customs and has talked with great nostalgia and longing for the times she and her women friends went to the mountains to camp out and gather century plants.

When the crowns were fully baked and cooled the charred leaves were pulled off and discarded; the unburned leaves and any center portion were then eaten or put in the sun to dry for future use. When needed, the leaves were soaked and the pulp from the inner side was scraped, much as we eat artichokes. The leaves could also be boiled with water to make a syrup. Sometimes the fibrous leaves were simply chewed, the sweet pulp being swallowed and the fiber quid expectorated. Ground mescal was also mixed with water to make a beverage.

Cliff Wood, an Arizona rancher whose family ranch is in an area that used to be inhabited by Aravaipa Apaches, remembers the Apache mothers dipping a quid of agave fiber in meat broth or stew and giving it to small babies to suck on.

Some tribes, such as the Chiricahua Apaches always gathered century plants in the late spring or early summer, collecting only those plants whose centers were beginning to sprout the tall flower-bearing stalk. The plants not blooming were known as "man" plants and those bearing an emerging flower stalk were called "woman" plants. They believed that the "man" plants were bitter. Daisy Johnson has remarked that it does not make any difference. I have baked both the "man" and "woman" plants with good results—although the "woman" plants would tend to be sweeter, having stored up energy (sugar) in preparation for flowering.

The young, growing mescal stalks were also roasted in pits or over coals. The Havasupais, who favored roasted stalks, would place a heavy rock on the bud so that the stalk would grow into a round lump rather than a tall stalk.[3]

Plants that were not harvested but allowed to flower also produced some foods, although they are of minor importance. A sweetish liquid was shaken from the flowers of a century plant after a rain. The fluid was emptied into a cup and drunk. The flowers were also collected and boiled. I tried some and found them a bit musty-tasting, but edible. The seeds were sometimes ground for flour.[4]

Among some tribes, such as the Mescalero and Chiricahua Apaches, an alcoholic drink was prepared from the roasted agave crowns. The fresh-roasted pulp was pounded and placed in a leather pouch and then buried for two days. When the pulp was removed from the ground the juice was squeezed out and allowed to ferment for another two or three days. The White Mountain Apaches would bake the crowns for 15 days, producing a semigelatinous mass that was crushed, the juice then collected and allowed to ferment.

In Mexico the much larger species of agave that grows there is cultivated commercially to produce the alcoholic drinks pulque, mescal, and tequila.

People have told me that prepared agave tastes like brown sugar, like pineapple, or like candy. I think it has a taste all its own—a pleasant, sweet flavor. The century plant is nutritious in addition to being tasty. A study has shown that a $\frac{1}{4}$-cup serving of prepared agave provides thirty calories and more calcium than does half a glass of milk.[5]

Raw agave is poisonous. Do not try even one nibble of this plant in the uncooked state or your mouth may be badly burned.

The following recipe is for baking an agave heart in your oven at home and some ways to use it.

BAKED AGAVE

Agave palmerii and *A. parryi* have fairly fleshy leaves. Use only these types as the other leaves are much too caustic. Because the raw juice of even these two types are very irritating to my skin, I equip myself with gloves, a long-sleeved shirt, long pants, and shoes before I begin hacking away at them. Indians used a sharp stone to cut off the leaves, but a steel machete-type knife works much better. Starting at the bottom of the plant, cut off the long ends of the leaves about where they begin turning white. The fastest way is to use strong, firm slashes, but be sure to keep the fingers of your other hand out of the way. When you have cut off all the leaves you should have a white-ish heart the size of a very large cauliflower. Rinse any accumulated dirt off the heart with water. Wrap the crown in aluminum foil and bake at 350° for about 10 hours, more if the heart is very large, depending on the size of the agave. When baking, place the foil-wrapped succulent in a shallow baking pan, because as the heart cooks it begins to give off juice, which is better in the pan than on the bottom of your oven. The cooked heart is soft and mushy, and golden brown.

Extract the fresh pulp by starting with the largest, juiciest leaves. Place a leaf on a cutting board and run a dull knife down the leaf in the same direction as the fibers, pressing to force out the pulp. The bottoms of the young center leaves will be almost all fiber-less pulp. The very center of the crown is pure pulp, Scoop this out, too. A medium-size agave will yield about 3½ cups of juice and pulp. The pulp is now ready to use in agave nutbutter or black walnut punch. (See "Black Walnuts," page 55.)

At this point the Indians dried any of the mescal that they wished to store. Sometimes the drying leaves were sprinkled with unfermented agave juice to give them a glaze and help preserve them.

AGAVE SYRUP

Use freshly baked leaves or well-soaked dry leaves. Place in a heavy pot with a close-fitting lid and cover with water. Simmer the leaves in the water for 2 hours, occasionally mashing the leaves with a potato masher to force out the juice and pulp. After 2 hours, remove the pan from the heat, and lift the leaves from the pot. Put the solid matter through a food mill or wring the leaves with your hands to extract the remaining pulp and juice, which should be combined with the juice in the pot. Discard the fiber. Boil the juice down to a thick dark brown syrup. Use the syrup for sweetening mush or in baking breads.

AGAVE NUTBUTTER

Mix ground sunflower seeds, pinyon nuts, or black walnuts with baked, fresh mescal pulp that has been separated from the fiber. The proportion is 2 parts ground nuts to 1 part mescal pulp. A dollop of honey turns these nutbutters into a delicious breakfast spread that can be used on cattail pollen muffins, saguaro seed bread, or mesquite bread.

This recipe for Agave Chiffon Pie was invented by two Tucson biology teachers to be served at an annual wild foods party for their students, and has been described as "not very sweet but quite tasty."

5

AGAVE CHIFFON PIE
Yield: 1 9-inch pie

3 to 4 cups mashed agave juice
2 eggs separated
1 cup sugar
¾ cup milk
1 tablespoon unflavored gelatin, softened in ¼ cup hot water
½ teaspoon salt
1 envelope low-calorie whipped topping, whipped
1 teaspoon pumpkin pie spice, nutmeg, or cinnamon
1 teaspoon vanilla
1 9-inch baked pie shell
¼ cup sliced almonds

To make agave juice, split a 3-to-4-foot plant, remove the blades, cut the heart in 8 to 10 1-inch pieces and put in a large pot. Cover with water and boil, uncovered, 3 to 5 hours. Remove from the heat and mash with a potato masher until it is thick and soupy; squeeze out the juice.

Beat yolks until thick. Combine agave juice, ½ cup sugar, milk, and yolks in top of double boiler and cook over boiling water for 10 minutes, stirring constantly. Remove from heat and add the softened gelatin. Mix to combine and chill until thick, about 1 hour.

Beat egg whites until stiff, gradually adding the remaining ½ cup sugar. Fold together the chilled agave mixture and the whipped topping. Then fold in egg whites, spices, and vanilla; pour the mixture into baked pie shell. Top with almonds and chill until set, about 1 hour.

OTHER USES: The long fibers in the leaves were used to make rope, carrying nets, sandals, cradle mats for babies, and stuffing for recreational balls.[6] If the spine on the end of the leaves was left attached to several long fibers, it could be used as a needle and thread.

A light brown paint was obtained from the hardened mescal juice, which covered the pit stones after a baking. This was used as a cosmetic by young girls who put it on their cheeks. Sometimes Apache scouts also used this paint on their faces, and it was occasionally used to paint stripes on buckskin.[7]

Southeastern Yavapai warriors made armor from plates of cooked, pounded, molded mescal. Mescal juice was used as waterproofing.

Apaches used a section of the flower stalk as the base for a fiddle and they used the dried, baked leaves with the fibers fluffed out as hairbrushes for people and horses.[8] (I saw the same type of brush on sale in a Mexican market for use as a pot scrubber.)

In the Hopi villages the stalk was placed on top of the Agave clan kiva in November to indicate that initiation into the one-horned or Agave society was in progress, and both the stalk and fiber were used in preparation of ceremonial equipment.[9]

Medicinally, agave also had several uses. The younger leaves were sometimes chewed as a tonic, which may have been a method of replacing a vitamin lack.[10] If signs of scurvy had already developed, mescal juice was the accepted antiscorbutic remedy. The leaves were roasted in hot ashes and pressed to yield juice. The juice was cooked and skimmed. A small amount was drunk by the patient in the mornings on an empty stomach.

Pfefferkorn, a Jesuit priest who traveled in the Sonora Desert in the mid-eighteenth century, wrote of this remedy: "The drink is uncommonly bitter and bad tasting but it completely cures the evil in a few days."[11]

Compresses were made out of wet macerated pulp and were used on local infections or bound on the chest to relieve congestion. Also, the juice of the root was applied to fresh wounds with a saturated cloth. If bleeding resulted after childbirth, a cloth soaked in mescal wine was introduced as far as possible into the birth canal.

barrel cactus

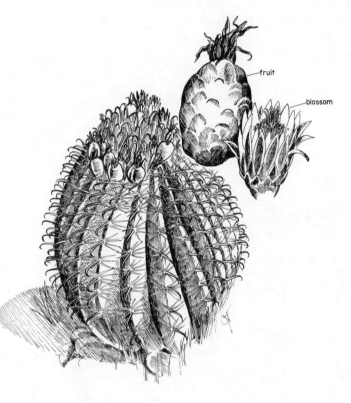

fruit

blossom

ALSO CALLED: Visnaga, devil's head, and compass cactus.

SCIENTIFIC NAME: *Ferocactus wislizenii* and other species.

HABITAT and DESCRIPTION: Barrel cactus is heavy-bodied with a simple cylindric shape. The various species of cacti range from 2 to 10 feet in height. The heavy spines of the cactus usually have hooked ends. Barrel cacti are found in southern Utah, southern Nevada, Arizona, New Mexico, West Texas, and Mexico. The yellow or orange flowers and the resulting fruits form a circle on the crown of the plant.

Barrel cacti are well-known as plants that are supposed to provide life-sustaining liquid to a person lost on the desert and dying of thirst. While containing some liquid, the barrel does not have a center filled with cool, refreshing water. To get at the juice, which might be quite bitter, the traveler must hack through the heavy spines to cut a slice off the top of the plant. The inner pulp of the plant can then be mashed or squeezed to yield a liquid. The oxalic acid in the plant tends to make some people ill and might cause nausea in a seriously dehydrated person, thus in fact worsening his plight.

Taller plants tend to grow faster on the shaded side, making them lean toward the south, hence the name "compass cactus."[1]

The fruit of the barrel cactus is yellow. It has a sour, but not unpleasant taste, and can be eaten directly from the plant. Cactus candy is made from the pulp of this plant. The recipe below comes from the Pima

County Cooperative Extension Service. In this modern recipe sugar is used for sweetening; the Indians had a method of making cactus candy without sugar. Small pieces of pulp were placed on top of dry mesquite beans in a pot with a little water and slowly boiled for a long time.[2]

BARREL CACTUS CANDY
Yield: about 3 dozen pieces

To prepare cactus:
Select small barrel cactus. With long sharp knife remove spines and outer layers. Cut the pulp crosswise in 1-inch-thick slices. Soak the slices overnight in cold water. Remove from water, cut in 1-inch cubes, and cook in boiling water until tender, about 1 hour. Drain.

To prepare syrup:
3 cups sugar
1 cup water
2 tablespoons orange juice
1 tablespoon lemon juice
Powdered or granulated sugar for garnish

Heat all ingredients together until sugar is dissolved.
Measure 2 quarts of cactus cubes. Add cactus to syrup and cook until nearly all of the syrup is absorbed, being careful not to scorch. Keep heat low and stir occasionally with a wooden spoon to keep candy from sticking. Remove cactus from syrup. Drain and roll in granulated or powdered sugar. For colored candy, any vegetable coloring may be added to the syrup.

OTHER USES: The hooked thorns were often made into fishhooks by the Pimas.[3]

9

cholla

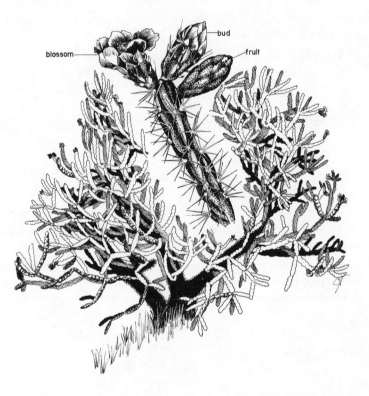

blossom • bud • fruit

ALSO CALLED: Staghorn and jumping cactus.

SCIENTIFIC NAMES: *Opuntia fulgida, O. acanthocarpa, O. echinocarpa, O. whipplei,* and *O. versicolor.*

HABITAT and DESCRIPTION: The cholla is a very spiny cactus with cylindric joints. The brightly colored flowers range from yellow through orange and from red to magenta. The species above are found up to 4,000-foot elevation in southwestern Colorado, southwestern Utah, southern Nevada, southern California, western New Mexico, and Arizona.

Indians sometimes called early spring—March—"the cactus moon" because food was scarce, and this plant was often the only available vegetable food.[1]

The Hopis, Maricopas, Pimas, Papagos and other tribes often used the cholla. The parts most eaten were the buds, which were picked before they opened into flowers in April and May. Also used were the fruits, which ripened in late summer, and in times of scarcity, the joints themselves.

The buds, fruit, and joints were traditionally pit-baked before drying for storage. A pit was dug and a fire built in it. Large stones were put on the fire to heat. When the fire died down the stones were removed and a layer of saltbush (*Sueda arborescens*) was placed over the hot coals. A layer of cactus fruit or buds was added next, then hot stones, and so on, alternating layers of hot stones and cactus to the top, ending with a thick layer of salt-

10

bush and topped with earth. The cacti baked overnight and in the morning were spread out to dry in the sun.[2] Buds that were baked or steamed before drying were less prone to spoilage.

This method of cooking required little water but consumed quantities of human energy. Indians today still use cholla buds, but they prefer to boil them instead of pit-baking.

Cleaning the thorns from the buds is a tricky job. The Hopis cleaned the thorns from the buds by putting them in a yucca-sifter basket with several small pieces of sandstone. The buds and stones were stirred or shaken together until the thorns had been removed.[3] A Papago woman told me that she removed the thorns by rolling the buds around in a basket, using greasewood twigs as a brush, but neither of these methods worked very efficiently for me. I found the easiest way to clean the buds is to fill each of two saucepans one-third full of clean gravel. Next, add the cholla buds and pour the gravel and buds from one pan into the other four or five times. Then check the buds to see which of them are almost clean. The remaining thorns can be removed with your fingers or a pair of tweezers. Then continue pouring gravel and buds together until most of the stickers are removed by the gravel. When all the buds have been dethorned, wash under running water.

Like other green vegetables the calorie content of cholla buds is low; a 2-tablespoon serving contains 48 calories, but this small serving contains more calcium than does a glass of milk.[4]

CHOLLA BUDS—BASIC PREPARATION

After thorns have been removed using the above method, put cholla buds in a saucepan, add water, cover, and boil for 15 minutes or until tender. Drain. Cholla buds can be used as they are in green salads, potato salad, or meat stews.

To preserve, set in the sun on trays to dry. To freshen dried buds, soak in water for 3 hours. Boil for ½ hour.

CHOLLA BUDS AND SQUASH
Yield: 2 servings

½ cup cholla buds
1 yellow crookneck squash

Cook cholla buds, drain, and pat dry with a paper towel. Slice crosswise very thinly. Slice yellow crookneck squash crosswise in about ¼- to ½-inch slices. Heat a small amount of bacon grease or vegetable oil in frying pan. Add sliced squash and cholla buds and sauté uncovered until tender, about 20 minutes.

11

FRIED CHOLLA AND CORN MUSH
Yield: 2–3 servings

¼ cup parched cornmeal
¼ cup cholla bud meal (about 2 dozen dried cholla buds)
1⅓ cups boiling water
½ cup cold water
½ teaspoon honey, agave syrup (p. 5), or other natural
    sweetener
½ teaspoon salt
bacon grease, oil, or butter

Parch corn by toasting in a frying pan, then grind. Or slightly brown commercial cornmeal in a hot frying pan.

Grind dried cholla buds finely. It may take several runs through the food mill to get them fine.

Bring the 1⅓ cups water to a boil in a saucepan. Combine parched cornmeal, cholla bud meal, ½ cup cold water, and honey or other sweetening. Mix together and add slowly to the boiling water. Cook and stir until thick. Cover and cook over *low* heat for 10 minutes. Pour into a small flat pan and cool. Refrigerate, if possible, for a few hours. Mixture will become solid. Turn out onto a plate, slice, and fry in bacon grease, oil, or butter. Serve with butter or perhaps a vegetable sauce. (See "Greens," p. 103.)

OTHER USES: Baked buds were ground and made into gruel to be given to patients suffering from stomach trouble and needing a special diet.[5]

Among the Hopis,[6] the root was chewed raw or pounded, boiled, and the liquid drunk for diarrhea; and the Apaches said that boiled roots made the best laxative for babies and small children.[7] A tea made from the roots was also used as a hair tonic.

Gums of various types have been obtained from the cholla. Some of these gums are used in Mexico as a cloth-stiffener in local textile mills.[8]

blossoms

ocotillo

ALSO CALLED: Slimwood and coach whip.
SCIENTIFIC NAME: *Fouquieria splendens*.
HABITAT and DESCRIPTION: The ocotillo is a large, thorny shrub with long, whiplike, unbranched limbs originating at a central base. Showy bright red flowers occur in dense clusters at the ends of the stems. Ocotillos usually are found 5,000 feet or lower from western Texas, through New Mexico and Arizona, and into eastern California.

The ocotillo is a plant that is amazingly well adapted to the desert. For most of the year its grayish thorny, long tendrils—often up to 9 feet high —sway stoically in the desert winds. But as soon as moisture hits the ground, the plant, which gave the appearance of being dead, sends out small green leaves to begin manufacturing food for itself.

Late one spring, upon changing residence, I found an ocotillo in my new backyard. I built a small well around the base of the plant and watered it. In only two days several well-formed leaves appeared on the stalk. Within a week the plant was fully leafed out. But as soon as the water supply dries up, the leaves wither and fall off, thus eliminating the broad leaf surfaces that would soon dissipate the plant's hoarded supply of moisture.

Both the seeds and the flowers are reported eaten by the Cahuilla Indians, but when I tried the seeds they had an alumlike drying quality and made my mouth feel very strange. The flowers, however, make a very tasty punch.

13

OCOTILLO FLOWER PUNCH
Yield: amount of water in bucket

½ bucket (any size) of ocotillo flowers
½ bucket of water

Soak the ocotillo flowers in cool water overnight. In the morning strain off the liquid. Makes a delicate juice.

DESERT PUNCH
Yield: 3 pints punch

1 quart ocotillo punch (see above)
1 pint squawberry juice (see "Squawberry," p. 79)
honey to taste, or other natural sweetener
mint sprigs

Combine ocotillo punch and squawberry juice. Sweeten with honey. Garnish with mint sprigs.

Note: This can be combined with any other juices, wild or domestic. Try different combinations.

OTHER USES: The strong, lacy skeletons of dead ocotillo are often used as a building material along with adobe mud. Also, because of its hardiness, the long whips are often used in the Southwest to make living fences. The tendrils are just cut off the plant and stuck in the ground side by side. If watered enough they will take root and leaf out, forming a very attractive and impenetrable living fence.

The ocotillo was also used medicinally. Apaches relieved fatigue by bathing in a decoction or tea made of the roots of the ocotillo, and they also applied powdered root to painful swellings.[1]

Ignaz Pfefferkorn, a very early Spanish traveler to the Sonoran Desert, wrote this account of the medicinal use of ocotillo:

The hocotillo [sic] is a remedy in driving away falls, bumps or crushing. A Spaniard who, together with some Indians accompanied me on a journey, fell with his horse so his right leg lay under the animal. Because of the weight lying upon it, the leg was crushed and swollen to such a size in a few minutes that in order to lay it bare, the boot and stocking had to be cut open. I was very much distressed because of this accident, but an Indian consoled me with the promise of devising a cure on the spot. He lighted a fire immediately, cut off some twigs from the hocotillo and after peeling these, roasted the remainder for a short time in hot ashes. Then he pressed out the juice on a cloth and bound the leg with it. This treatment he repeated several times and in two hours' time the swelling was gone and the Spaniard was without the least pain.[2]

14

fruit

prickly pear

ALSO CALLED: Tuna, Indian fig (fruit) and beavertail (plant).

SCIENTIFIC NAME: *Opuntia phaeacantha* (formerly *O. Engelmannii*) and other species.

HABITAT and DESCRIPTION: This cactus is found in desert areas throughout the West. It is characterized by flat, fleshy joints or pads that grow in clumps. Large, waxy flowers appear in the spring, followed by juicy red fruits.

A Navajo legend says that when prickly pears are being gathered, a hair must be plucked from the gatherer's head so that the plant will yield fruit without twisting its heart.[1] This is typical of many of the myths connected with wild plant foods. The Indians were so very dependent on the foods they gathered that they did not dare risk offending the spirits that ruled the plants or the plants themselves, which many believed to have souls. To do so might mean starvation.

There is a belief among many tribes that eating too many tunas (fruits) give the eater chills or shivers. The first time I encountered this information I assumed that it was a folk custom, but after finding four or five other references giving the same information, I felt there might be basis for the warning. Pimas,[2] Yumans,[3] and Apaches[4] were among the tribes believing that too many prickly pears lead to illness. The Pimas thought it was the darker purplish fruit that was "poison." I have not been able to

15

disprove or substantiate this information because few people today eat many tunas at one time. Probably only someone lost on the desert without food would eat as many prickly pears as a hungry Indian who might have a whole meal or two of them.

None of the recipes here, eaten in normal quantities, is likely to make the average person sick. They should be good for you, because tunas are very high in calcium.[5]

The pads of the prickly pear, called nopales when prepared for eating, are rather mucilaginous. The plant transforms the water it absorbs into this thicker substance, forestalls evaporation and makes it easier to withstand the long periods of drought.

Some species of *Opuntia* occur in Europe also. An American Southwesterner, for example, traveling along the Mediterranean would be amazed to see prickly pears growing there. Botanists believe all the species have been introduced from the Americas.

PRICKLY PEAR FRUIT JUICE—Basic Preparation
Yield: 1 quart

Using tongs, collect the pears. Usually the darker red ones are the riper and juicier, but the color of the fruit varies from species to species. You can tell if a fruit is ripe if the pulp looks red at the place you have detached it from the plant. Resist the urge to do much tasting in the field unless you are very careful of the thorns. I have spent several uncomfortable afternoons with thorns in my gums from unwise munching. Once stuck in your mouth the stickers are irritating and impossible to see, although they do dissolve or become dislodged in seven or eight hours.

When you get home, holding the fruit with tongs, brush under running water with a stiff vegetable brush. Fill a medium-sized sauce pan with water and bring to a boil. Put six pears into the boiling water and blanch for about 10 seconds. Remove the pears from water with tongs. This method seems to render the thorns less irritating and makes it easier to handle and peel the fruit. Blanch and peel only six at a time because, when the fruit cools, the thorns again become stiff.

Peel the tunas and discard the peels. Slice tunas in half and extract the seeds. (Your thumb is a good tool to use for this.) Put the fruit in one bowl and the seeds in another. When you have about ¾ of a bowl of seeds, fill the bowl with water and, using your hands, break up the seed clusters so that the pulp clinging to the seeds will disperse in the water. Mash the other bowl of pure pulp with a potato masher and strain through a mesh strainer or a colander lined with cheese cloth. After the seeds have soaked for a couple of hours, strain off the accumulated liquid and add to the liquid that has drained from the mashed pulp. Combine in a saucepan and simmer for 5 minutes. Pour into clean glass jars and refrigerate. Two dozen tunas yield about 1 quart of juice by this method.

Prickly pear juice can be used to prepare jelly (begin recipe where

16

it says below, "To prepare jelly"), and can be mixed with other fruit juices for a tasty punch. Try it with 1 pint cactus juice, one pint cranberry juice, and 1 quart gingerale.

PRICKLY PEAR JELLY
Yield: 6 medium jars

To Prepare Fruit:

It will take at the minimum 1 quart of fruit pulp (14 to 20 fruits depending on size) to make the 2½ cups of juice you will need for the jelly recipe. Do not pick the fruit overripe unless you use a few underripe pears to add the needed pectin. Use tongs to gather the fruit. Rolling the fruit around in the grass will remove some of the stickers—the rest will come off when the fruit is pressed through the cheesecloth.

To clean the pears, brush them with a vegetable brush. Wash in running water and place in a kettle or covered saucepan with enough water to cover. Boil until tender and soft, about 20 minutes. Drain water from the fruit and mash with a potato masher.

Line a colander or mesh strainer with two thicknesses of cheesecloth. Strain mashed fruit. Set strained juice aside so the sediment can settle to the bottom. For a clear jelly, do not use this sediment.

To Prepare Jelly:
1 package (1 tablespoon) powdered pectin
2½ cups juice
3 tablespoons lemon or lime juice
3½ cups sugar

Combine pectin and juice in a pan. Stirring constantly, bring to a fast boil and add the lemon or lime juice and the sugar. Bring to a hard boil and boil for 3 minutes. Timing is important to get a good jell. Remove from the fire. Stir and skim off foam. Fill jars that have been sterilized and seal with hot paraffin.

The Pima County Cooperative Extension Service has collected the following modern recipes for prickly pears.

CACTUS PICKLE
Yield: 2 quarts

2 quarts prickly pears
2 cups sugar
⅔ cup vinegar
3 ounces red cinnamon candies
whole cloves, optional

17

Measure the 2 quarts of prickly pears whole. Remove the skins, cut each fruit in half lengthwise and remove the seeds. Cook pear halves until transparent in a syrup made of the sugar, vinegar, and cinnamon candies. If you use cloves, put them in a little cheesecloth bag so they can be removed before the pickles are put in jars that are equipped with standard canning lids. Process for 15 minutes in a boiling-water bath.

CACTUS DATE CONSERVE
Yield: 5–6 medium jars

2 cups thinly sliced prickly pear fruit (seeds removed)
1½ dozen dates, chopped and pitted
juice of 1 orange
grated rind from 1 orange
2 slices canned pineapple, cut into small pieces
4 teaspoons lemon juice
½ cup pineapple juice
1½ cups sugar
⅓ cup broken walnut meats

Combine all ingredients except the walnuts. Cook slowly in heavy pan until of desired jamlike consistency for conserve. About 5 minutes before removing from fire, add the nutmeats. Pack in sterile jars and seal with paraffin.

CACTUS PRESERVE
Yield: 5–6 medium jars

2 quarts prickly pears
1 slice orange, ¼ inch thick
1½ cups sugar
⅔ cup water
2½ tablespoons lemon juice

Measure the prickly pears whole. Peel, cut in halves lengthwise, remove the seeds. Cook the peeled, seeded tunas and the orange slice in a syrup made from the sugar, water, and lemon juice until transparent. Remove the orange before packing the preserve in sterile jars. Seal with paraffin.

NOPALES—Basic Preparation

Using tongs, collect the new young pads in the spring. Those from 1 inch to 3 inches in diameter will be the most tender. Place the pads in a saucepan and cover with water. Boil for 20 minutes. A large clove of garlic and slice of onion will add flavor (Hopis often add an ear of sweet corn for flavoring.) Drain. To clean the pads, place on a hard surface (plate or cutting board), and using a small, sharp knife, scrape off the stickers and the rubbery leaves. (There are so many thorns on the edge of the pads that I usually just trim away the edge about 1/16 inch). If you are not working with very small pads, the part where the joint was attached to the rest of the plant is apt to be tough; cut it off. Rinse the pad well under water. A strong light, either natural or artificial, is necessary in order to see if you have removed all the thorns. Scrutinize each pad carefully. It is better to get a thorn in your finger now than one in your tongue or lip later. Rinse the plate or cutting board before cleaning the next pad or you will "re-sticker" the new pad as you work. Chop or dice the cleaned pads or cut them into thin strips of about ½ inch. These can be covered with water and stored in the refrigerator for several weeks.

Mexican Indians sprinkle the chopped nopales over their everyday bowl of beans to add crunchiness and the taste of something green to their monotonous diet. Nopales can also be added to a tossed salad or stirred into any casserole.

The southwestern Indians who lived within the confines of what is now the United States usually ate the nopales boiled and unadorned. These recipes are Mexican in origin.

If you find the above instructions too tedious, you can buy canned nopales in supermarkets in the Southwest. Canned nopales or homemade nopales stored in water should be rinsed well before using. Either the canned or home-prepared variety can be used in the following recipes.

NOPALES AND EGGS—(Nopaleggs)

Yield: 2 servings

2 tablespoons finely chopped onion
1 tablespoon oil or butter
2 tablespoons diced nopales
4 eggs, beaten
1 tablespoon water
Salt and pepper

In a heavy pan fry onion in oil or butter until transparent. Add nopales and heat thoroughly. Beat eggs and water. Reduce heat to low, and after frying pan has cooled slightly, add eggs and scramble. Season to taste.

CACTUS CONDIMENT
Yield: ¾ cup

¼ cup chopped onion—Spanish or green or a combination
Oil or bacon grease
¼ cup nopales
3 tablespoons red chili paste
½ cup water

Fry onion in oil or bacon grease until transparent. Add nopales and fry for about 1 minute. Add red chili paste and water; simmer until nopales are well-saturated with sauce. Use as a sauce for meat.

BEAVER-TAIL STEW
Yield: 2¼ cups

1 cup nopales
¼ cup chopped onion
Oil
1 teaspoon salt
2 cloves garlic, minced finely
1 large tomato or 2 small tomatoes, chopped
1 to 2 tablespoons chili paste
2 shakes of comino (cumin)
½ cup cooked shredded pork, beef, or venison

Fry the nopales and onion in oil until they are slightly crisp. Add garlic, tomato, chili paste, comino, salt, and meat. Simmer until the tomato is done. If the tomato was not very juicy, add water, tomato juice, or gravy so the mixture can simmer without burning. Serve with toasted tortillas.

SWEET AND SOUR NOPALES
Yield: 2¼ cups

3 strips bacon
2 cups nopales, sliced about ¼ inch thick and 2 or 3 inches long
2 or 3 tablespoons chopped onion
3 tablespoons vinegar
2 tablespoons sugar

Cut bacon into small pieces and fry. When crisp remove from fat and drain. Fry chopped onion in bacon fat until transparent. Add nopales, vinegar, and sugar, and stir over low heat until flavors are combined.

OTHER USES: Indians of various tribes often bound wounds with
the soaked split pads. The Pimas placed heated nopale pads on the breasts
of women with new babies to encourage the flow of mother's milk.[6]

In New Mexico the roasted split pads are bound on the neck and
chin to reduce swellings caused by mumps, and in Baja California the pads
were used to abate the swellings from rheumatism.[7]

The mucilaginous juice of the pads has been used by the Mexicans
to make mortar.[8]

Although the prickly pear tends to dominate lands that are over-
grazed, taking water and space from any browse plants that might grow
back, it does have its compensations: During drought cattlemen singe the
thorns off the pads to make forage for their animals. Cattle seem to like
the fruit also.

saguaro

fruit

ALSO CALLED: Giant cactus and sahuaro.

SCIENTIFIC NAME: *Cereus giganteus* (formerly *Carnegia giganteus*).

HABITAT and DESCRIPTION: This huge succulent grows only in the well-drained desert soils of southern Arizona and Sonora, Mexico, and in a small area in southeastern California near the Colorado River. The highest recorded altitude at which it has been found is 5,100 feet but generally it does not grow above 3,500 feet. The largest plants attain a height of more than 50 feet and develop as many as 50 arms. Some plants are believed to be 150 to 200 years old.

Cereus giganteus is not only great in stature but also in importance in the lives of the Indians. The harvest of the juicy, crimson fruit of the giant saguaro was so important to both the Pima and Papago Indians of southern Arizona that it signaled the beginning of the new calendar year for both tribes.

The saguaro flowers appear in May and the fleshy fruits ripen five to six weeks later. At harvest time in June, entire villages would move out into the saguaro forests and set up temporary camps for two to three weeks. The event was an occasion of great jollity and preparation for the yearly wine drinking ceremony, which was believed to bring the rain. It was also a time to lay in a large store of food for the months ahead. There was much gorging: the saguaro was almost the first fresh food of the year and the first taste of anything sweet.

22

One colonial priest, writing in the seventeenth century, reported that some Indians became so corpulent after eating huge quantities of the sugar-rich fruit that he was sometimes unable to recognize at first sight individuals otherwise perfectly familiar to him—and this was after feeding for about three weeks on the fruit.[1] This is understandable, as each fruit contains about 34 calories (two tablespoons of dried saguaro seed have 74 calories), and is also high in protein, fat and vitamin C.[2]

Early morning in the Papago saguaro camps would find everyone out gathering the fruit. To knock the fruit off the top of the tall plants they used two saguaro ribs spliced together with a short traverse stick of creosote bush or catclaw affixed on the end.

If the fruit did not split open on hitting the ground, the gatherers split the fruit with the sharp edges of the circular calyx attached to the fruit stalk. The pulp was scooped out from each half of the rind with two motions of the thumbs and thrown into a basket, while the shell was thrown to the ground, care being taken that the red lining fell uppermost, for this hastened the rain.

At midday, when it became too hot for gathering, the morning's harvest was taken back to camp, where the pulp was soaked in water to loosen the minute black seeds that were then sun-dried, to be ground and used for bread or mush. The remaining pulp was further prepared for syrup or jam. The mother or mother-in-law always officiated at this occasion. Whenever a large clay jar or olla was filled, it was sealed air-tight with a potsherd (piece of broken pottery) cemented over the opening.

At the end of the harvest it was time for the great celebration: much drinking of saguaro wine accompanied by dancing, singing, and speech-making. For the wine-making, every family contributed a jar of boiled

Saguaro pulp and seeds are spread out to dry in the foreground of this scene of a typical Papago saguaro fruit–gathering camp. The large clay jugs, or ollas, are used to store the juice from the saguaro fruit. (*Arizona State Museum— University of Arizona Helga Teiwes, photographer.*)

saguaro juice to the large jars ceremonially guarded in the council house. As soon as the juice was poured from its air-tight container it was mixed with four times the amount of water—the ideal being a mild intoxicant that could be taken in quantity. A small fire was lighted in the council house to keep up a steady, moderate heat and official tasters directed the fermentation process; the usual length of time was about 72 hours. The wine, called navai't, is almost impossible to keep, so traditionally the entire supply had to be consumed in 24 hours. The drink was sort of a cider (considered by some to have an unpleasant taste) which, when drunk in ritual quantities, induced vomiting. Women did not become intoxicated but looked after the men when they did.[3]

The Maricopas, a central Arizona tribe, also welcomed the time of the saguaro harvest, not because the fruit formed a considerable portion of their staple foods, but as an occasion for celebration and debauchery. The giant cactus did not grow in the Maricopa territory proper—they had to travel a little way to get it. Some of the fruit was eaten fresh in the field, but the main point was to collect as much juice as possible. Sometimes as many as a hundred large pots of juice were set in the meeting house to ferment. Each family owned from one to three of the potfuls which they were called to carry home when ready.

Spier writes:

> Friends in other tribes were invited to dance at this time, by saying "Our storehouse is ripe." The guests came in a group, camping at some distance from the village. The messenger who bore the invitations now walked out to the camp where he named those invited by each Maricopa, pointing out where each lived.
> The method of handing out drinks was stereotyped. The men invited would bring three or four friends. The host would send his wife into the house to bring out a cupful of wine which he gave to his guest. He continued plying him with liquor until he was drunk. Then he himself drank and after that he furnished [wine] to the friends of the man invited.[4]

The song for this dance was sung at no other time. It told of blood and battles, and the intoxication and incitement of the song usually ended in a decision to go on a raid. The Pimas and Papagos, however, ended their party by recovering from their hangovers and packing up the food they had collected to take back to their permanent villages. Often they pressed the fruit into large cakes that they dried for future use, but they could not keep these long as they became wormy, as I have learned only too well from experience!

After reading that the Papagos had had trouble preserving their fruit, I resolved to be especially antiseptic while preparing the fruit I had harvested. I formed the fresh fruit into small bars on a tray lined with waxed paper. I put them to dry in a screened-in porch, where sunlight but no flies could get to them. When the bars were dry I sanitarily wrapped each one in aluminum foil and stored the lot in a coffee can with a plastic lid. I paid little attention to them for about a month and a half until one day a new and most interesting beau who was very interested in anthropology came to

24

visit me for the first time. To impress him with my engrossing field work, I told him about my saguaro-harvesting activities and trotted out a sample of the candy. When he unwrapped the foil he gave the dried fruit bar such a funny look that I thought he must be queasy about trying something new and different.

"Go ahead and eat it," I urged. "It's really good."

"But . . . ," he stammered, "it has some things on it. They're little white worms."

With horror, I saw that my carefully preserved specimen was crawling with worms. Apparently the eggs had been laid in the fruit before I gathered it. Heat destroys the eggs, but I had not cooked the pulp.

My own experience with saguaro-gathering now includes several harvests. A word of advice to the enthusiastic neophyte gatherer: go out as early in the morning as you can. Walking around the desert, up and down the arroyos, and poking at the fruits require some exertion, and it is very hot during the season when the fruits ripen. Sometimes it is so warm that you could swear it must be noon but it is really only 8:30 A.M. or so.

Saguaro-gathering seems to work best in pairs: one person knocks the fruit off the cactus while the other person catches it as it falls. It is important to try to catch the fruit as it falls, especially ripe fruit, for it will burst open upon hitting the ground and pick up gravel that is almost impossible to clean out. I use the frame from an old trout net fitted with a shallow muslin bag for a "catcher." A small plastic pail also works, but the net frame is easier to use because it has a handle. I imagine an old tennis racquet minus strings and with a bag of some sort attached would do just as well.

The Indians' idea of slitting the fruit with the sharp edge of the calyx and scooping out the pulp with the thumbs works as well now as it did then, but a small knife can be used instead. Do not forget to take a bucket to put the pulp in.

PREPARING THE FRUIT FOR THE SYRUP

When you get home with your harvest, immediately combine the pulp with water. Use water equal to the amount of fruit you have. Stir the mixture well, using your fingers to break up the clumps as much as possible. Then let it stand for a day. A clean dish towel over the bucket helps keep flies away.

The next day, using a fine wire-mesh strainer, strain all the liquid—which will now be gloriously red—into a large pot. Add a little more water to the remaining seeds and pulp and soak again for several hours. Then strain off that juice into the same pot. Boil the juice until it is reduced to at least half the original quantity or less if you want a stronger syrup. Store in the refrigerator in clean covered jars.

PREPARING THE SEEDS

Spread the seeds remaining from the above syrup-preparation process on a large flat pan or tray and dry them in the sun. There will be some washed-out pulp adhering to them. When the seed mass is dry, break up the clumps of seeds and whitish pulp. Then, using a pan with sides at least two inches high, vigorously shake the pan back and forth. After a few shakes you will notice the whitish waste coming to the top while the shiny black seeds settle to the bottom. Skim off the white material with a spoon and discard it. Continue the shaking and skimming process until you are left with mostly smooth shiny seeds and just a little white pulp. (A small amount of this pulp left with the seeds will not hurt.) Store the seeds in a can or jar with a tightly fitting lid.

Molly Manuel, an energetic Papago woman living on the San Xavier Papago Reservation, near Tucson, spent several afternoons teaching me to cook with saguaro. She gave me the following recipes for saguaro pudding and saguaro candy.

SAGUARO PUDDING
Yield: approximately 2 quarts

1 cup whole wheat flour
6 cups of water
1 cup finely ground saguaro seeds
2 teaspoons salt

In a small bowl mix 1 cup flour with 1 cup water. Bring the remaining 5 cups of water to boil in a large saucepan. Add the flour and water mixture to the boiling water. Reduce heat and cook and stir until the mixture is thick. Add the ground saguaro seeds and salt. Cook and stir a few minutes longer.

The pudding can be eaten with sugar and cream as a breakfast cereal. If a bit less water is used the pudding can be made thick enough to be used as a side dish for dinner with gravy or butter. For the Indian of long ago, of course, such a pudding *was* dinner.

BEEBEROS   (Saguaro Seed Candy)
Yield: 16 nut-sized candies

2 cups finely ground saguaro seeds
6 or 7 tablespoons saguaro syrup

Grind saguaro seeds into a fine meal with a food mill or electric blender. Measure them into a bowl. Add saguaro syrup, beginning with 4 tablespoons, mix and then add another tablespoon and mix. The exact amount of syrup to be added will depend on the amount of moisture in the seeds and on how thick you have made your saguaro syrup. The mixture should be moist and slightly sticky but should hold its shape when rolled into walnut-sized balls. Put on a tray covered with waxed paper to dry.

The typical aboriginal Indian bread was made without leavening and was reported to be heavy and indigestible. One loaf—only three inches thick and twenty inches in diameter—might weigh fourteen pounds.[5] The following is a modern version of saguaro seed bread, using baking powder.

SAGUARO SEED BREAD
Yield: 1 6-inch diameter loaf

⅔ cup ground saguaro seed
⅓ cup ground sunflower seed
¾ cup whole wheat flour
2 teaspoons baking powder
sprinkling of salt
1 tablespoon honey
2 tablespoons saguaro juice or syrup
½ cup water

Combine the ground seeds and flour, baking powder and salt in a mixing bowl. Add honey and liquids. Form into a ball-shaped loaf on a greased baking sheet and bake at 350° for 35 minutes. Serve with butter and jam.

Variation: For a fruitier bread, use more saguaro juice and cut down correspondingly on the water.

Saguaro juice can be used to make the following:

FRUITS OF THE DESERT SALAD

1 tablespoon unflavored gelatin
¼ cup cold water
¾ cup boiling water
3 tablespoons frozen orange juice concentrate
1 cup saguaro juice

Soften the gelatin in the cold water for 3 minutes. Add the boiling water and stir until the gelatin is dissolved. Add the frozen orange juice concentrate and the saguaro juice. Refrigerate until thickened to almost solid. Stir in your choice—or a combination—of orange sections, peeled seeded prickly pears, and black-walnut kernels.

Saguaro juice can be combined with other juices for a fruit punch— a little goes a long way. Or, it can be used as a topping for ice cream.

The following recipes for saguaro cactus jam and jelly were contributed by the Pima County Cooperative Extension Service.

SAGUARO CACTUS JAM
Yield: approximately 2–3 pints

Half cover 4 cups saguaro cactus pulp with water and soak for 1 hour, stirring occasionally. Boil in a covered pan over a low flame for ½ hour. Strain off the liquid, reserve the pulp, and boil the liquid slowly to a syrup, stirring constantly because it burns easily. Crush the pulp and put it through a fine sieve to remove the seeds. Add the pulp to the thickened syrup and cook to the consistency of jam. The jam is made without sugar; the fruit contains enough of its own sweetening.

SAGUARO CACTUS JELLY
Yield: 3 pints

Gather the seedy centers of the ripe saguaro cactus fruit. Place in saucepan and add water to about 1 inch or a little more above the seedy mass, cover, and simmer for about 20 minutes. Put 3¼ cups of this juice in a large kettle and add 1 package powdered pectin and ¼ cup lemon juice.

Bring to a full boil, then add 4½ cups sugar all at once. Bring to a boil, stirring all the while, and boil 1 minute.

Pour into jelly glasses and cover with paraffin.

OTHER USES: Saguaro pudding was given to women who had just had babies "to keep the stomach warm and make the milk flow."[6]

The saguaro seeds, which contain substantial quantities of vitamin C, were fed to chickens. The seeds were also used as a medium for tanning.

Ribs of the dead cacti were used as splints for broken bones and were bound together for shelves and to make carrying baskets.[7]

yucca: datil

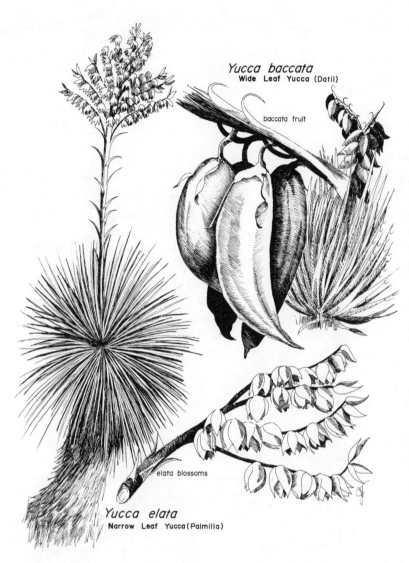

Yucca baccata
Wide Leaf Yucca (Datil)

baccata fruit

elata blossoms

Yucca elata
Narrow Leaf Yucca (Palmilla)

ALSO CALLED: Banana yucca, blue yucca,
Spanish bayonet, and wide leaf yucca.
SCIENTIFIC NAMES: *Yucca baccata* and
Y. arizonica.
HABITAT and DESCRIPTION: The datil is
found in southwestern Colorado, southern
Utah and Nevada, and southwestern Texas to
southeastern California. It grows on
mesas and foothills at elevations of 3,000 to
8,000 feet, often with pinyon and juniper.
Y. arizonica grows below 4,000 feet in
southern Arizona and northern Mexico. These
yuccas usually have no leafy stem, the
crowns sitting level with the ground. The
flower stems are usually short, while the
fruits are quite long and banana-shaped,
turning reddish when ripe. Fruits appear from
May to September, depending on the
elevation.

yucca: palmilla

ALSO CALLED: Soap tree, Our Lord's Candle,
narrow leaf yucca.
SCIENTIFIC NAME: *Yucca elata.*
HABITAT and DESCRIPTION: The palmilla
yucca is found in the grasslands and deserts
of western Texas to central and southern
Arizona at elevations of 1,500 to 6,000 feet.
Yucca elata has a prominent stem covered
with dried, straw-colored leaves and one
or more crowns of spine-tipped bright green
leaves. The clusters of white flowers are
borne on the ends of long stalks that appear
in the spring.

The various species of yucca are considered by many to have been the most
important of the wild plants utilized by the Indians of the Southwest. The
plant occupies this ranking because of the wide variety of uses to which it
could be put and to the accessibility of the genus throughout the Southwest.[1]

Besides serving utilitarian uses from food to fiber, yucca was also
used ceremonially. The roots of the yucca have a high component of saponin,
which has detergent properties. When pounded and soaked in water, the
roots, called amole, form copious suds. The suds were used for washing hair
and garments, and ritually in a great many ceremonies.

The Hopis and Tewas at Hano use yucca suds, which depict clouds,
at all ceremonies of adoption and name-giving. When an infant is named
before sunrise on the twentieth day after birth, its head is washed by the
paternal grandmother and each member of the father's clan who gives an
additional name smears the child's head with suds.[2] Among the northeastern

29

Yavapais, both the mother and her newborn child were washed with yucca suds.[3]

Yucca was also used in the wedding ceremonies of the Hopis at Oraibi with both the bride and groom having their hair washed by their respective future mothers-in-law. The hair of a deceased Hopi was washed with yucca suds and tied up with yucca fiber. For this reason, it was considered dangerous to dream about the yucca root or headwashing as it was believed to foretell death.[4]

In the northern Arizona villages of both the Havasupais and Walpais the girl's puberty rites included a bath and a shampoo with yucca suds. Yavapai and Walapai warriors purified themselves with yucca baths when returning from battle.[5] Navajo, Ute, and Apache scalps, when they were brought to Hano pueblo, were washed before sunrise with amole.[6]

The flowers and stem of the palmilla were eaten by Indians. The large waxy white flowers were picked in the spring and boiled and seasoned. The stem was collected any time from March to late summer, peeled, and baked just as agave. (See directions under "Agave," p. 4.) After the stalk had been baked, it was dried in the sun. When needed, pieces of the dried stalk were soaked in water and the sweet liquid was strained and drunk.[7] The dried ground flowers were considered a delicacy by the Navajos, who roasted them and used them for thickening soups. The hard capsular fruits of the *Yucca elata* were not eaten.

Large fleshy fruits appear on the *Yucca baccata* and they were the most relished part of that plant. The fruits were eaten raw, baked, boiled, dried, and ground into meal. The tender central leaves were also cooked in soups, boiled with meat, and used in a variety of culinary combinations.

The Zuni made a social occasion of yucca fruit preparation: During the day, the women of the household and their friends worked, boiling the fruits and peeling them when they had cooled. The pared fruit was heaped in large bowls. At night the group was joined by male friends and relatives.

Everybody sat on the floor surrounded by the filled and empty bowls. The fruits were seeded and chewed, after which the chewed fruit was ejected from the mouth into one of the empty receptacles. When the work was finished the hostess served a supper. The next day the chewed fruit was boiled. When it was the right consistency it was cooled and made into pats about three inches in diameter. These pats were dried on a roof in the sun for about three days, then several dried pats were combined to be formed into long rolls about the size of a rolled section of newspaper. These rolls were further semi-dried on the roof and then stored. The dried fruit was eaten by breaking a piece from the roll and eating it dry, or a section was soaked in water and made into a syrup.[8]

Among the Apaches, the banana-shaped fruits were gathered before fully ripe and laid on twigs covered with greens to ripen in the sun; when ripe, the fruit was roasted in hot ashes. The skin on the roasted fruit turned black and was easily stripped off. The baked pulp was spread on yucca leaves to dry for two days in the sun, the drying yucca being adorned with fresh sunflower blossoms to make it attractive. The sunflower was used as a sym-

30

bol of the importance of the sun to the growing of all things and this use constitutcd a prayer that the sun continue to make the yield plentiful year after year.[9]

Papagos also used the seeds from the fruit of the datil, which they ground into meal.

The recipes for Yucca Flower Soup and Yucca Hash came from Mexico, where the yucca flower also is used. While plain boiled yucca blossoms might not appeal to everyone, these two recipes are very tasty. The flowers should be gathered only from the palmilla.

YUCCA FLOWER SOUP
Yield: 5 cups soup

3 cups petals from *Yucca elata*
2 tablespoons butter
1 tablespoon cornstarch
1 quart very rich milk, or milk combined with cream
salt
pepper
dash of nutmeg

Use only the petals of the flowers. Discard the centers of the flowers, as they are very bitter. Wash the petals in a colander. Put in saucepan and cover with water. Boil for 15 minutes. Drain and return to pan. Mash the petals with a petal or food masher. Add butter and cornstarch and cook for 2 minutes. Slowly add milk and cook until mixture has thickened. Add nutmeg, salt, and pepper, seasoning to taste.

YUCCA HASH
Yield: 3 cups

2 cups petals of *Yucca elata*
1 tomato
1 onion
3 cloves garlic
½ green pepper
1 tablespoon sugar
½ cup boiled sliced cholla buds or 1 small can peas
salt
pepper

Use only the petals of the flowers, as the heart is very bitter. Boil the petals in two cups of water for 15 minutes. Drain. Dice the tomato, onion, garlic, and green pepper, and simmer in a small saucepan until tender. Add sugar, salt, and pepper to taste. Combine sliced cholla buds or peas, yucca petals, and other vegetables, and heat thoroughly. Serve on fried tortillas.

PREPARATION OF BANANA YUCCA

Boil fruit for 20 minutes to ½ hour. Drain fruit, cool, peel, and seed. Mash pulp, return to pan, and cook until of desired consistency for jam. Fruit may be further sweetened with a little honey. If thickened with a little flour it makes a good filling for turnovers and it can be used in other pastries.

The cooked pulp can be spread in a thin layer and dried in the sun or in a very slow oven, of about 200°. The resulting sheet can be rolled or folded and stored. It is good for snacks and to take as a quick-energy food on hikes.

If the fruit pulp is put into a sieve or lined colander and allowed to drain, the resulting juice may be boiled down into a syrup. The ripe fruits may also be seeded and sliced and used as a substitute for apples in a pie.[10]

OTHER USES: Yucca fibers were used in all phases of Indian life: for sandals, baskets, matting, cordage, fabric, fishing and carrying nets, head rings—which women used to aid them in balancing pots on their heads, straps, and cradle boards. Loosely woven yucca formed a base for feather cloth and fur cloth. The fibers were also used in brooms and brushes for hair, paint, dusting, and pot scrubbing. Dolls and a variety of games were fashioned from the leaves.[11]

Medicinally yucca fibers were used to ease many illnesses. A patient with a cold, rheumatism, or a wound would crush yucca leaves into fibers. Then he would enter the sweathouse where he induced vomiting to expel the poisons by inhaling large quantities of steam and chewing the leaf fibers.[12]

At Hano, a northern Arizona Tewa village, small ceremonial bows of cedar are strung with yucca and in some male initiation ceremonies there, the novices are beaten with yucca whips. Neckties of knotted yucca strips are worn by clowns during some dances.[13]

The juice of the yucca fruit is used by the Hopis as a varnish on certain kachinas.[14]

During World War I, yucca fiber from Texas and New Mexico was used to make 80 million pounds of bagging and burlap, but this manufacture was not continued after the war. During World War II, when jute fibers were again in short supply, a factory was built at Kingman, Arizona, to extract yucca fiber. The failure of this venture was blamed on poor planning and a lack of information on how to separate the fiber from the other leaf material.[15]

II
Nuts and Seeds

acorn

SCIENTIFIC NAMES: *Quercus emoryii, Q. reticulata, Q. turbinella, Q. grisea,* and *Q. arizonica.*

HABITAT and DESCRIPTION: These oak species (except for *Q. turbinella*) may grow to become large trees up to 60 feet high or they may mature as shrubs. The different species will be found in the 3,000- to 8,000-foot altitude range in the foothills and canyons of western Texas, Colorado, New Mexico, Arizona, California, and northern Mexico. They produce a small, relatively sweet acorn, which ripens in late July and August.

With all the southwestern Indians doing their food-gathering in supermarkets these days, it is only those native dishes with special appeal that have survived time to appear regularly on the dinner table. Acorn stew is one that has not faded in popularity. Many Apache housewives still keep a store of acorn meal on hand to make this much relished dish. Some collect and use more than one hundred pounds of acorns a year.

According to Grace Mitchell, leader of the Yavapai tribe, Yavapai cowboys who work on the desert carry only a pocketful of acorns and some water for lunch. "They say it really fills them up," she remarked.

Other southwestern Indians who used acorns include the Pimas, the Papagos, and the Navajos.

34

The Indians of central, northern, and coastal California used acorns to a much greater extent than did the desert Indians. To the California Indians, acorns were the staple and most important food. The type of acorn that grows in that area of California is much larger than the desert variety and also more bitter, owing to a greater amount of tannin. The tannin had to be leached out with water before the seeds were palatable.

Some of the southwestern Indian groups ate acorns only raw or ground into meal that was used to thicken stews. Other groups roasted the acorns before eating them.

When gathering acorns, reject any shells with a little hole, for that means a worm has gotten to that acorn before you. To further cull the acorns, put them all in a bucket of water; the hollow ones will float. A sprinkling of wood ashes will protect stored acorns against worms.

Apache women crack the acorn shells on a flat stone using a 6- or 7-inch stone cylinder, which they roll over the nuts. A regular metate works just as well, as would any two flat stones.

ACORN STEW
Yield: 3–4 servings

1 pound stewing beef
½ cup finely ground acorn meal
salt
pepper

Place beef in heavy pan and add water to cover. Put lid on pan and simmer beef until it is very tender and is almost falling apart. Remove the beef from the liquid and chop the meat into very fine pieces. Return meat to the liquid in the pot. Stir in the acorn meal. Add salt and pepper to taste. Heat the mixture and serve.

ACORN BREAD
Yield: 1 9-inch loaf

1 cup acorn meal
½ cup cornmeal
½ cup whole wheat flour
3 tablespoons salad oil
1 teaspoon salt
1 tablespoon baking powder
¼ cup honey
1 egg
1 cup milk

Shell acorns and grind meats in a food mill or electric blender. Measure 1 cup meal and combine with cornmeal, flour, salt, and baking powder. Combine honey, egg and milk and add to dry ingredients. Mix just until all dry ingredients are moistened. Pour into greased 8 x 8-inch pan and bake at 350° for 20 to 30 minutes.

OTHER USES: A tea made from the green bark of the oak is reportedly a powerful astringent, and antidiarrhetic. It was also used as a cure for bad-smelling feet.[1]

The wood is used for fuel and fence posts.

Sorghum halapense
Johnson grass

Oryzopsis hymenoides
Rice grass

grass seeds

ALSO CALLED: Johnson grass.
SCIENTIFIC NAME: *Sorghum halapense.*
HABITAT and DESCRIPTION: This perennial is found throughout the West (except in Washington, Idaho, and Montana) in fields and along irrigation ditches. It is a large plant with long leaf blades and terminal branching flower heads (which produce the seeds).

ALSO CALLED: Rice grass, Indian millet.
SCIENTIFIC NAME: *Oryzopsis hymenoides.*
HABITAT and DESCRIPTION: This grass is found in dry, open woods and sandy plains at medium elevations, throughout the western United States. The large seeds grow singly on the ends of the branched spikes. There are many stiff leaf blades.

The Indians collected and ate more than fifty different kinds of grass seeds. Johnson grass and rice grass are representative of the types that were used.

Some of the other grasses that were used include dropseed and giant dropseed (*Sporobolus cryptandrus* and *S. giganteus*), barnyard grass (*Echinochloa crusgalli*), and panic grass (*Panicum obtusum* and *P. bulbosum*).

Because the seeds are generally so small and the harvest not extensive, no one type of grass seed was ever considered an important food. If enough seeds were found, they could be boiled for mush or made into bread, but usually the seeds were combined with other flours and ground meal to add variety to the standard fare of breads and mushes made with cornmeal and mesquite flour. Sometimes the ground seeds were used to thicken meat gravy.

One Indian woman described grass-seed collecting in this way: "We would work all day and feel as though we had done so much and there would be about a cupful [of seeds]."[1]

37

Seed-gathering was usually the task of the women. Some Indian groups used seed-beaters to gather the seeds and some just used their hands to strip the seedheads into a basket held beneath the plant or into a receptacle suspended in front by a cord passing around the neck. Some of the seeds with tough chaff and hulls were lightly ground on a metate or pounded on a wooden mortar before being winnowed by tossing into the air from a pottery pan or a hide.[2] Sometimes another method was used: if the seeds were very vigorously shaken in a cloth bag the seeds would settle to the bottom and the chaff would rise to the top.[3]

I have tried mixing water with grass seeds that have been slightly ground on a metate. The chaff, hulls, and seeds with hulls generally float and can be skimmed off while the completely clean seeds sink to the bottom.

All grass seeds can be eaten raw but usually they taste better when dried, roasted and ground.

OTHER USES: Various fresh grasses were used as insulation for foods cooked in baking pits or on coals. Layers of clean grass were also used as mats on which to dry foods in the sun.

nut

jojoba

ALSO CALLED: Goat nut, deer nut, wild hazel, and coffee bush.

SCIENTIFIC NAME: *Simmondsia chinensis.*

HABITAT and DESCRIPTION: The jojoba is found in southern Arizona, southern California, Sonora, and Baja California, or the area known as the Sonora Desert. It grows at elevations between 1,000 and 5,000 feet and is found in well-drained, coarse desert soils on dry slopes and washes. The incidence varies from a few plants to 200 per acre. The bush is an evergreen shrub with thick, leathery leaves, and is seldom more than 6 feet high. The acornlike fruit has from 1 to 3 seeds in a capsule and ripens from mid-July to mid-September, depending on the elevation and the climate.

Although Indians have been munching on jojoba nuts for centuries, it is not as a food that this fruit is being investigated today. Quite a bit of attention has been focused on this little bush since the discovery that oil of the jojoba bean—chemically a liquid wax—makes a good substitute for sperm whale oil, and industrially is also useful in many other ways. The fruits are from 48 to 51 percent oil.

The University of Arizona Office of Arid Lands Research sponsored an international conference bringing together experts in all phases of jojoba utilization. If it develops that there is a large commercial demand for jojoba nuts, it is hoped that Indian groups that inhabit areas of heavy jojoba growth will be interested in harvesting the natural crop as a means of generating income on their reservations. This has already been attempted, to a limited extent, and now there is a stock of jojoba nuts available to researchers who might find a use for this wax.

Experiments are now being conducted with jojoba oil in such divergent areas as an ingredient in an ointment for acne prevention and as an agent to retard evaporation of water stored in open reservoirs.

The one study undertaken on jojoba meal as an animal feed (using the part of the bean left after the oil was extracted) was not successful. Although the entire jojoba bean was found to have a protein content of 35 percent, the study also showed that the oil is 80 percent indigestible.

The Indians, however, have used jojoba nuts for years with no apparent ill effects, but never considered them a staple or consumed them in great quantity, just as an occasional snack.

Raw, the nut has a bitter taste due to tannin.

The jojoba bush is extremely well-adapted to desert conditions. Lapsing into leafless dormancy, it can withstand one or two years of rainless weather, rebudding from old wood when rain does come.

An outdoorsman friend suggested the following method for roasting jojoba nuts. The recipe for the beverage comes from the southern California Cahuilla Indians and was modified somewhat by early pioneers.

PEACOCK'S ROASTED JOJOBA NUTS

Shell jojoba nuts and place on a shallow pan or tray (such as a jelly-roll pan.) Roast for 1 hour at 250°, stirring once or twice. Take out of the oven and cool for a few minutes. Put the nuts in a bowl and add just a few drops of vegetable oil over the nuts, toss lightly, and sprinkle with salt. (The additional oil helps the salt adhere to the individual beans.) These are very tasty.

JOJOBA COFFEE
Yield: 1 cup

Method 1: Grind roasted jojoba nuts in food grinder. Combine 1 heaping tablespoon ground jojoba nuts, the finely crumbled yolk of a hard-boiled egg, and 1 cup of water. Stir until combined. Boil for 3 minutes. Strain into a cup. Add sugar and cream to taste.

Method 2: Same as Method 1 except omit the hard-boiled egg yolk and boil the mixture for 3–5 minutes. Strain into a cup and add cream and sugar.

OTHER USES: In 1763 an unknown Jesuit priest wrote that he found the Indians of the area using the jojoba medicinally. The Indians, when wounded by an arrow, would place 2 or more jojobas—as many as there was room for—in the wound. Reportedly, the swelling was prevented until they were able to further care for the wound.[1]

Ignaz Pfefferkorn wrote that it was considered a good remedy for stomachache "especially in cases where the stomach has been chilled." The oil also was used in cases where a swelling became hardened because of cold

40

"and would not open or resolve itself"—the oil being spread on the swelling.[2]

Both the nuts and the leaves themselves make a nutritious browse for animals, which can eat it faster than it can grow.

Jojoba oil is used cosmetically in hair oil, shampoo, and body oil. These products are widely available in Mexican pharmacies.

mesquite

ALSO CALLED: Honey mesquite and pechit.
SCIENTIFIC NAME: *Prosopis juliaflora.*
LOCATION and DESCRIPTION: The mesquite
is a shrub or small tree growing to 25 feet,
with a compound leaf and many small
leaflets. The greenish-yellow flowers are long
and cylindric and the fruits are long and
podlike. Mesquite trees grow at 5,000 feet
or below from southern Kansas to southeastern California and northern Mexico. It is
common along washes and wherever it can
get enough water. When the beans are fully
ripe they are brittle and straw-colored. Some
varieties sport beans that are randomly
streaked with red.

To the Indians of the Sonora Desert, mesquite beans were their most
important food, even more important than corn. It was mesquite atole
(gruel) that sustained them day after day through the winter until the next
cultivated crop was ready. Yumans, Mohaves, Cocopas, Pimas, and Papagos
agreed that no other wild or cultivated food compared in importance with
the mesquite bean and its close relative the screw bean (*Prosopis pubescens*).[1]
The Seri Indians of Sonora have eight different words to describe mesquite
beans in every state from a tiny pod of less than 1 inch to the fully ripe
and dried pod that has fallen from the tree onto the ground.[2] The Pueblo
Indians at Acoma and Laguna also used the beans and the Yavapais accepted
them in trade.

The mesquite bean provides goodly amounts of protein, carbohydrate,
and calcium. Four tablespoons of mesquite meal provide seventy calories.[3]

For a Maricopa woman, ability to gather and prepare mesquite beans

42

was the major accomplishment of a good housewife, for skill in this task would keep her family from hunger. A large part of the woman's day was spent pounding and grinding mesquite beans.

Every industrious family built and filled a cylindrical granary bin to hold their mesquite harvest. The women would go out every day in groups —for protection against the enemy—and collect beans until the trees were completely stripped.[4]

The ripe beans were dried on the rooftops before being stored. Those beans that were not considered good enough to be stored whole were ground into meal and stored. The meal was sprinkled with water and formed into small, round, hard cakes. These cakes were used on damp days when the whole beans had absorbed too much water to be easily ground.[5] Slices of the cake were cut off and fried like mush, used to thicken gravy in a stew, or eaten raw. The flour was used in breads and beverages. A fizzy, slightly alcoholic drink was made by fermenting a mixture of mesquite and water.

The old Papagos still look upon the mesquite bean with respect when they reminisce about earlier days. The aged aunt of my Papago informant, Molly Manuel, listened to the two of us talk about uses of the mesquite beans for several hours. Finally she said quietly, "The Indians ate good food. They never sickened and they got real old."

The following recipe for mesquite gruel was given to me by Molly Manuel. The Papagos ate only two meals a day. During the winter mesquite gruel is what they probably ate at both meals. It was often varied by the addition of various grass seeds, pieces of animal fat, and sauces made of dried greens or berries.

MESQUITE GRUEL
Yield: approximately 3½ cups

3 cups water
⅔ cup finely ground and sifted mesquite flour
⅓ cup whole wheat flour
½ teaspoon salt

Combine 3 cups of water and the mesquite flour in a saucepan. Boil together for 10 minutes. Mix the whole wheat flour with ½ cup of water. Add to the hot mixture. Simmer, stirring occasionally, until thick.

MESQUITE BROTH
Yield: 2–3 cups

Wash and break up about 4 cups of fully ripe, dried mesquite beans. Put them in a heavy saucepan and cover with water. Boil for 2 hours, checking the beans occasionally and adding water when necessary. The beans and the liquid will turn reddish. Mash the beans several times with a potato masher or heavy spoon while they are cooking. Remove the beans from the water and cool. Reserve liquid. Wring and tear the beans by hand or put them through a food mill to help separate and loosen the pulp from the fiber. Return the whole mass to the liquid and simmer, tightly covered, for ½ hour. Strain off the thickened liquid, and discard the fiber and seeds. The broth is now ready to use in any of the following authentic Indian dishes.

MOLLY'S MESQUITE DUMPLINGS
Yield: 4 dumplings

2 cups mesquite broth
½ cup whole wheat flour
¼ cup water
1 teaspoon vegetable oil
sprinkling of salt

Mix the flour, water, vegetable oil, and salt together, stirring and kneading until it forms a fairly elastic ball. Divide the dough into 4 parts, forming each part into a small, flat tortilla. Bring the mesquite broth to a boil in a pan with a tightly fitting lid. Slide the dumplings on top of the liquid, reduce heat, and cover. Cook for ½ hour. What you should have is rather hard dumplings with a thick brown sauce.

Variation: The above recipe produces a rather plain food that might not be to the taste of modern desert-dwellers. It can be improved by adding 1 teaspoon of baking powder and ¼ cup raisins to the dough.

MESQUITE PUDDING
Yield: 4 servings

1½ cups mesquite broth
2 tablespoons whole wheat flour

Mix the flour with a little of the cool broth. Add the mixture to the rest of the broth and bring to a boil. Cook and stir until pudding is thick; it will thicken more as it cools.

Variation: This makes a very good dessert acceptable to most modern palates by adding 2 tablespoons brown sugar or honey, ¼ teaspoon cinnamon, a sprinkling of nutmeg, and ½ cup raisins.

Anthropologist and professor Thomas Hinton watched this mesquite beverage being prepared by descendants of the Opata and Pima Indians in rural Sonora, Mexico.[6]

ATOLE DE PECHITA
Yield: 1 cup

1 cup mesquite broth
1 tablespoon brown sugar or panocha (Mexican cone-shaped raw sugar)

⅛ teaspoon cinnamon
sprinkle of ground cloves

Combine broth, brown sugar, and spices in saucepan. Heat and stir until sugar is dissolved. Serve warm or chilled. It is very good.

MESQUITE MEAL (for use in recipes below)

Select sound beans with no worm holes. Pass the beans through a food grinder or pound and grind them on a metate. The seeds are extremely hard and will resist crushing. Using a sieve or a sifter basket, shake and sift the ground beans. The ground dried pulp of the beans will readily pass through the sifter. Discard the pieces of the outer shell and the seeds. Regrind the sifted meal if a finer product is desired.

MESQUITE BREAD #1
Yield: 1 7-inch round loaf

1 cup finely ground and sifted mesquite meal
1 cup whole wheat flour
1 teaspoon baking powder
1 teaspoon baking soda
2 tablespoons oil
¾ cup water

Combine dry ingredients. Add oil and water and mix until dough forms a ball and cleans the sides of the bowl. (Because mesquite beans and meal have a tendency to pick up any moisture from the atmosphere, the amount of water needed will vary with the weather.) Lightly grease a cookie sheet or flat pan. Form the dough into a half-sphere loaf on the pan. Bake for 30 minutes at 350°.

The above recipe makes a very earthy loaf and the flavor of mesquite is quite pronounced. The less gastronomically adventurous might try the following recipe.

MESQUITE BREAD #2

To any yeasted bread recipe calling for about 8 cups of flour, substitute from ½–2 cups of mesquite meal and proceed as usual. This will make a darker, somewhat heavier bread than your regular recipe would produce.

MESQUITE BOILED BREAD
Yield: 10 bread balls

½ cup water or more
½ cup finely ground and sifted mesquite flour
½ cup whole wheat flour
½ teaspoon salt

45

Combine the water, flour, and salt to make a medium stiff dough. The amount of water needed will be governed in part by how much moisture is in the wheat and mesquite flour. Bring a large pot of water to a boil. Form dough into balls about the size of a large walnut. Drop each dough ball into boiling water and boil for about 15 minutes. Remove from water with a slotted spoon. This bread is very good served with meat gravy.

MESQUITE PUNCH
Yield: 1 serving

2 tablespoons fine mesquite flour
1 cup water

Combine mesquite flour and water. Stir and let sit for 2 minutes. Strain. This mixture was also used as a dip for tortillas and fry bread. It was considered quite refreshing on a hot day.
Variation: Mix with other wild juices.

OTHER USES: The Pimas made a hair dye by boiling the black gum in water and applying the mixture to the hair with a rag, after which they plastered mud over the hair and left it on all night. To finish the treatment, the hair had to be washed off in three tubs of water the next morning before sunrise. This same resin was used to mend pottery, or when boiled in just a small amount of water, it was used as a paint for pottery.[7]

Basketry materials were obtained from the inner bark of the mesquite bush. The hard wood makes excellent firewood and fence posts.

As a medicine the mesquite tree was almost a pharmacy. The black gum was boiled and the dilute liquid was used as a wash for sore eyes and open wounds.[8] It was also used for sore lips, chapped and cracked fingers, and as a lotion for "bad disease" (venereal disease) and sunburn. It was also used as a soothing throat gargle.[9] A mixture of powdered mesquite gum and finely strained sand was applied to the navel of newborn babies to prevent infection.

An emetic and cathartic was made from the liquid in which the inner bark had been boiled,[10] while the liquor from boiled dried mesquite beans was used to soothe severe sunburn.

The leaves were also used medicinally to make a tea good for headache and stomach trouble. The tea was placed directly on the eyes to cure pink eye, was held in the mouth to heal painful gums, and was swallowed to cleanse the system.[11] Because mesquite was considered a cool plant, it was never given for high fever.[12]

The mesquite tree served not only the human animals that lived on the desert but other animals as well. The foliage and beans are eaten by livestock and an excellent honey is made from its flowers by bees.

cone

nut

pinyon pine

ALSO CALLED: Nut pine and pignolia.
SCIENTIFIC NAMES: *Pinus edulis,*
P. monophylla, and *P. cembroides.*
HABITAT and DESCRIPTION: Pinyon pines
have short trunks and conic crowns. The old
bark is grayish-brown. They are usually
smaller than other pine trees. Nut pines are
found at elevations of 4,000 to 7,500 feet
throughout the West in large stands or mixed
with juniper trees.

At Santa Clara pueblo, pinyon is said to be the oldest tree and its nuts the oldest food of the people. It was the result of going up on the western mesa and eating the fallen pinyon nuts that the people "first knew north and west and south and east."[1]

The nuts are picked in the fall or early winter. Most of the tribes used them, often making long treks to the closest mountains to get them. The Havasupais, for example, climbed up to the edge of their isolated canyon to get them. The Navajos apparently collected the most and those tribes having a trading relationship with the Navajos could obtain extra supplies from them.

The seeds are very rich and supply considerable protein and fat, and have more than 3,000 calories a pound! Some Apaches did not permit their pregnant women to eat pinyons lest the unborn child grow too fat, making delivery difficult.[2] Pinyon gruel was, however, considered fine food for the baby after it was born.

47

Pinyon nuts were roasted to help preserve them. Minnie Marshall, a kindly and helpful Indian woman living in the village of Havasu, deep in that paradisiacal blue-water canyon, revealed that the nuts are parched by putting them in a basket with live coals. The basket and its contents are shaken so that the coals never burn the basket, but the nuts are well-roasted.

If you would rather not risk scorching your Indian basket, try roasting the nuts in a slow oven (250°) for 1 hour, stirring them occasionally. To shell them easily, while they are still warm, put them between two damp cloths and roll them vigorously with a rolling pin. To roast pinyons over an open fire, put them in a cast-iron skillet and shake them over the fire until a sample nut is toasted to your taste.

Besides being eaten raw or roasted directly from the shell, pinyon nuts were boiled into a gruel, ground and formed into cakes, or rolled into balls and eaten as a delicacy.

HAVASUPAI PINYON NUT BABY FOOD
Yield: ½ cup

¼ cup raw or roasted pinyon nuts (shelled)
¼ cup water
1 tablespoon honey

Grind nuts very fine. Combine with water in a small saucepan. Cover and boil for 5 minutes. Add honey. Modern adaptation: Double the above recipe. Add whatever quantity baby doesn't eat to regular pancake batter and cook pancakes for the rest of the family.

ZUNI PINYON NUT SUNFLOWER CAKES
Yield: 8 cakes

½ cup raw sunflower seeds (shelled)
½ cup raw pinyon nuts (shelled)
sprinkle of salt

Grind seeds and nuts together. Add salt. The mixture will be very rich and oily. Form flat little cakes, using about 2 tablespoons of the mixture for each cake. The Zuni wrapped the cakes in leaves and baked them in ashes; you can wrap each cake in aluminum foil and bake at 350° for about 50 minutes. Don't let the cakes get too brown or the delicate flavor will be lost.

Pinyon cakes were eaten with other food in the place of meat. They can also be crumbled into a weak stew or vegetable dish to add richness and body. You can use them in any number of imaginative ways and combinations.

Here are two recipes to start you off:

48

PINYON NUTS and SQUASH

Peel and cut into small pieces any yellow-meat fall squash, such as butter-nut squash. Steam until tender. Mash. Add crumbled pinyon nut sunflower cakes to taste.

PINYON SOUP
Yield: between 1 and 2 quarts

1 turkey or chicken carcass
1 carrot
½ onion
1 stalk celery
1 clove garlic
peppercorns
pinyon nut sunflower cakes

This is a good use for a turkey or chicken carcass after you have eaten most of the meat. Put the carcass, vegetables, and spices into a large pot and cover with water. Simmer for about 2 or 3 hours until the soup looks strong. Strain the soup through a sieve. If there are any bits of meat left on the bones, add them to the broth. Put broth in refrigerator until fat forms a solid layer, and can be easily removed. For each cup of broth add 3 tablespoons crumbled pinyon nut sunflower cakes. Salt to taste. Simmer for a few minutes. Watch closely as soup tends to foam and boil over. Serve hot in bowls garnished with chopped parsley or cilantro (coriander).

SWEET PINYON MUFFINS
Yield: 6 muffins

1 cup ground pinyon nuts
½ cup whole wheat flour
½ cup water
3 tablespoons honey
2 teaspoons baking powder
½ teaspoon salt

Combine dry ingredients. Add water and honey and mix well. Pour into greased muffin tins and bake for ½ hour at 350°.

OTHER USES: One Indian told me, "Pinyon nut meal is very greasy. In the olden days they used to grease shoes with it."

The pitch, from the pinyon pine, which was gathered in June, was used by the Hopis to waterproof and repair pots and by the Apaches to waterproof waterbaskets. Pitch was also used as chewing gum, and heated

49

pitch was sometimes applied to the face to remove facial hair and to protect cuts and sores from air.[3]

Among the Hopis the gum was put on hot coals and members of a mourning household smoked themselves and their clothes in the fumes after a funeral. During December, the gum was put on one's forehead when venturing outside the house, as a precaution against sorcery.[4]

Pinyon needles were used as a cure for syphilis by the Zunis. The patient chewed the needles, swallowed them, drank a quantity of cold water, and then ran for a mile or until he perspired profusely. When he returned home he wrapped himself in a heavy blanket. Women patients were not required to run.[5] A tea was made of twigs and drunk warm in conjunction with chewing the needles. Syphilitic ulcers were scraped with the fingernail until they bled and the powdered pinyon gum was sprinkled over them to promote healing.

seeds

sunflower

ALSO CALLED: Marigold of Peru.
SCIENTIFIC NAME. *Helianthus annuus.*
HABITAT and DESCRIPTION: The sunflower is a common and conspicuous roadside weed, which grows throughout the United States. The large flower heads have showy yellow petals and a yellow, brown, or purple-brown center. Stalks are 3–7 feet high. In the Southwest the blossoms are evident from March through October or November.

Sunflowers figure heavily in the myths of the southwestern Indians, and the symbol of the sunflower is often used in art and decoration. According to a Hopi legend, when the sunflowers are numerous, it is a sign that there will be an abundant harvest.[1] Hopi women ground dried petals and mixed the resulting powder with yellow cornmeal, using it to decorate their faces for the Basket Dance.[2]

A scalp song at Hano pueblo describes sunflowers as "watered by the tears shed by Navajo girls.[3] Anthropologist Matilda Cox Stevenson reported that in Zuni dances each personator of the rainmakers wore the symbol of the sunflower attached to the forelock.[4]

Sunflower seeds were the Southwest's only domesticated food plant before the introduction of beans, squash, and corn from Mexico. The plants needed little attention, for they grew rapidly in spite of weeds. The rich,

51

tasty meats of the wild sunflower were eaten both raw and roasted and ground into meal by nearly all Indians.

In the fall, the heads of the ripened sunflowers, whether cultivated or wild, were broken off, dried, and beaten with sticks to release the seeds. The seeds were winnowed, parched, and ground into meal on the metate, and eaten as atole (mush).[5]

Sunflowers are extremely nutritious. The experimental laboratories of the University of Illinois report that the seeds contain 50 to 55 percent protein, which is almost totally assimilable by the body. The seeds are also a rich source of vitamin B.[6]

The seeds contain about 50 percent oil, which the Indians extracted by boiling. The oil was used for cooking and for grooming the hair.

A Yavapai tribal leader says that her people used to roast sunflower seeds on coals for a few seconds, shelled them, ground them, and shaped the meal into an oblong cake. "We ate it just like that," she said. "It was sort of like peanut butter."

Besides general use as a thickener for soup and stews, here are some recipes for preparing sunflower seeds.

ZUNI SUNFLOWER PUDDING
Yield: 6 servings

1 cup corn kernels (from green, or very young, cobs)
1 cup finely ground raw or roasted sunflower seeds (shelled)
1 cup finely chopped squash (zucchini or yellow crookneck)
2 cups water
1 teaspoon salt

Put corn kernels through food grinder retaining all the juice or "milk." Combine with sunflower meal and chopped squash in a saucepan. Add water and salt. Cover pot tightly and simmer for 1 hour, stirring occasionally. Toward the end of the cooking period, remove cover, if necessary, so that the mixture can thicken. It should be thick and gelatinous. This dish is good warm or cold.

Variation: Near the end of the cooking, add chopped, roasted, peeled green chilis.

SUNFLOWER GRAVY
Yield: 2 cups

¼ cup tiny pieces of suet or finely chopped bacon
6 tablespoons fine sunflower meal
1 tablespoon cornstarch*

* Corn flour or wheat flour would traditionally be used but this guarantees a better result.

52

2 cups water
pinch of salt
2 to 3 tablespoons finely chopped onion

Fry suet pieces or bacon and onion until onion is translucent but not beginning to burn. Add sunflower meal, cornstarch, and salt and cook for a minute, stirring and watching so that it does not burn. Slowly add water, while stirring. Lower heat and cook until thick. Add more water if necessary (the amount needed depending on water content of other ingredients, which often varies in the arid Southwest).

Use as a sauce for vegetables or mush.

The following is a modern adaptation of Indian sunflower bread:

SUNFLOWER BREAD
Yield: 1 loaf

¼ cup honey
¼ cup soft butter
2 eggs, beaten
1 cup whole wheat flour
1 teaspoon salt
1 tablespoon baking powder
1½ cups ground sunflower seeds, shell and meat, or meats only
1 cup milk
½ cup whole or coarsely chopped sunflower meats

Beat together honey and butter. Beat in eggs. Combine flour, baking powder and salt and ground seeds. Add to honey/butter mixture alternately with the milk. Fold in whole sunflower meats. Pour into greased loaf pan and bake 1 hour at 325°. Cool on rack. This bread slices better when cool.

SUNFLOWER SEED COFFEE

Brown empty hulls of sunflowers in small frying pan. Watch them carefully so that they do not burn. Grind the browned hulls finely. For each cup of beverage, steep 1 teaspoon (or more to taste) ground hulls in 1 cup boiling water for 3 minutes. Drink plain or sweetened with honey.

(See other uses for sunflower seed under the "Saguaro" and "Pinyon" sections.)

OTHER USES: Hopis made yellow dye from the flower petals and black and purple dyes from the seeds. The dye was used for coloring baskets and cloth and for painting their bodies in certain ceremonies.[7] The Pimas and Maricopas of the Salt River Reservation used the inner pulp of the stalk as chewing gum. That same pulp, broken into pieces and strung on a string, made very fast-burning candles.[8]

The seed is very good for chicken feed, causing poultry to lay more eggs, while bees relish the flower nectar for making delicious honey.

Various parts of the sunflower were also used medicinally. Pimas made a bitter decoction or extract of the leaves, strained the liquid, and gave a tablespoon or more for high fever until the fever cooled. The same brew was applied to horses' sores caused by screw worm.[9]

The Zunis combined sunflower roots with two other herbs to cure rattlesnake bites. The theurgist, or medicine man, sucked the wound to extract as much poison as possible, then chewed the three roots together and with a bandage applied the masticated mass on top of the bite. This was repeated every morning for five days, when a woman aide of the theurgist washed the patient's head with yucca suds and both prayed for his recovery.[10]

Also, a root decoction was used as a warm wash for rheumatism.[11]

husk

shell

black walnut

ALSO CALLED: American walnut.
SCIENTIFIC NAME: *Juglans rupestris*, var. *major*.
HABITAT and DESCRIPTION: These native nut trees are found at elevations of 3,500 feet to 7,000 feet along streams and washes in Arizona, New Mexico, and Mexico, and in warm areas of California. They are commonly found near cottonwood and sycamore trees. The tree sometimes grows as high as 50 feet and provides good shade. The nutmeats are enclosed in a thick shell that is covered with a pulpy husk.

The town of Nogales, on the Arizona–Mexico border, has taken its name from the Spanish name for walnuts, *nogal*.

Because of the thick shell, very little meat is found in each nut, but what is there is quite tasty and very rich in fat—sometimes as much as 75 percent of the nut is oil.[1]

The soft husk surrounding the shell will stain the skin brown, so use gloves when husking walnuts. The hard nuts are best opened by banging them with a hammer, then using a nutpick or other sharp thin instrument to pick out the meats.

The Apaches especially relished the black walnut; they ate it raw, and also used it in cooking and baking. The following recipes were Apache favorites.

BLACK WALNUT PUNCH
Yield: 1 portion

2 tablespoons finely ground black walnuts
2 tablespoons prepared agave pulp (See "Agave" p. 4)
½ cup water

Combine ingredients in a jar, cover with a lid and shake vigorously until all the ingredients are blended. For added richness, add 1 teaspoon of honey.

WALNUT CORN BREAD
Yield: bread for 2—1 small loaf

6 ears fresh green corn
pinch of salt
½ cup coarsely chopped or finely ground black walnuts

Peel corn, saving husks. Cover husks or store in plastic bags to keep them from drying out. With a sharp knife cut kernels off corn. Put the kernels through a food grinder or blender, saving all the milky juice. Line a small metal bread pan with the corn husks, reserving some husks for the top. Combine the ground corn, corn juice, salt, and the ground or chopped walnuts. (Apaches made it both ways.) Pour mixture on top of corn husks. Cover with some of the remaining corn husks. Bake at 350° for about 1 hour. Since the juiciness of green corn varies, be sure to check the bread occasionally to prevent overbaking.

OTHER USES: The White Mountain Apaches used the juice of walnut hulls "like sheep dip" to rid horses and cattle of lice and other parasites. The same juice was considered good for cleaning maggots from wounds, and 1 or 2 tablespoons were often given to a dog for worms.[2]

The hulls, containing a strong and long-lasting dark brown dye, were rubbed on gray hair to make it dark again.

Sometimes a decoction of the walnut bark is used as a bath to ease the aches of rheumatism and pains in the legs.[3]

III

Grapes, Berries, and Cherries

chokecherry

ALSO CALLED: Wild cherry and stone fruit.
SCIENTIFIC NAME: *Prunus serotina*,
var. *virens*.
HABITAT and DESCRIPTION: The chokecherry is found from Canada to Georgia, in
New Mexico, Arizona, and California generally
at elevations of 4,500 to 8,000 feet, often
in or near coniferous forests. The tree, which
sometimes reaches a height of 25 feet, bears
small, sour fruits that vary from dark red
to almost black when mature. Each cherry
has one large bony seed. The fruit is ripe in
July and August.

Although we now have many ways of storing foods for use out of season, the Indians usually had to rely on just one method—drying. This is what they did when they found more chokecherries than they cared to eat fresh.

The Jicarilla Apaches ground the berries and made the meal into round cakes approximately 6 inches in diameter and 1 inch thick. These hard, blackish patties could be stored and reconstituted when they were needed by soaking in water. The seeds contain a fair percentage of cyanide, but this poison is volatile and driven off by cooking. The soaked cherry cakes were boiled and the juice strained and sometimes sweetened for use as a beverage, or the juice was combined with other ingredients.

Tellus Goodmorning, a wise, humorous man and a member of the Taos Pueblo Council, told me a story about his use for chokecherries:

58

I made chokecherry wine once. I crushed the berries and put in sugar. The bubbles came up. Then I put it in bottles but the caps kept popping off. But I put the caps back on. I was going to save the wine for Christmas time. But one night my son, who was a teen-ager then, and his friends took every bottle and went down by the creek and drank it. Their grandmother lived in the pueblo near there and she saw them lying drunk and asleep and told us what had happened.

Maybe life in that ancient pueblo is not too different from life in town.

An old-time western recipe for chokecherry wine, which I have not tried and do not guarantee, appears in the book *Cooking in Wyoming, 1890–1965*. It calls for washing, pitting, and putting the fruit through a food chopper. Then measure fruit and juice, and add the same amount of water. Let the mixture stand for 3 weeks. Stir daily. Drain off the liquid and add half as much sugar as liquid. Let stand until it stops fermenting—about six weeks—and bottle.[1]

The following recipe combining the ubiquitous corn meal with wild cherries has been adapted for modern tastes and cooking methods.

CHERRY CORN BREAD
Yield: 1 9-inch round loaf

To prepare cherries:
1 cup whole chokecherries
2 cups water
2 tablespoons honey

Place cherries and water in saucepan. Bring to a boil and cook for 20 minutes. Drain juice from cherries and reserve. Cherries can now be easily pitted—just give each cherry a little squeeze and the pit will pop right out. Combine pitted cherries and honey.

To prepare bread:
1 cup cornmeal
1 cup whole wheat flour
2 teaspoons baking powder
1 teaspoon salt
1 egg
1 tablespoon honey
2 tablespoons shortening or butter
juice from 1 cup cooked cherries

Heat oven to 425°. Grease a 9-inch frying pan and put in oven to heat. Combine dry ingredients. Add egg, honey, and shortening or butter. Measure cherry liquid and add water to make ¾ cup. Add to other ingredients. Stir in sweetened cherries. Pour batter into hot pan and bake at 425° for 20 to 25 minutes.

Some Indian groups made a cherry jam by cooking together wild cherries and honey. The following is a modern recipe for cherry jelly, given to me by a resident of Prescott, Arizona.

CHOKECHERRY JELLY
Yield: 5 pounds jelly

3 cups chokecherry juice (from about 3½ pounds ripe cherries)
¼ teaspoon almond extract (optional)
6½ cups sugar
1 bottle liquid pectin

Stem the cherries. To prepare the juice bring 3 cups water to a boil and add fully ripe cherries. Simmer, covered, 15 minutes. Pit the cherries, working over a pan or bowl so as to not lose any juice. Place cherries and liquid in a jelly bag and squeeze out the juice. Measure 3 cups of juice into a very large saucepan. If desired, add almond flavoring. Add the sugar to the measured juice and mix well. Place over high heat and bring to a boil, stirring constantly. Stir in pectin at once. Bring to a full rolling boil and boil hard for 1 minute, stirring constantly. Remove from heat, skim off foam with metal spoon, and pour into glasses immediately. Cover at once with ⅛-inch hot paraffin. Makes about 5 pounds of jelly.

OTHER USES: Chokecherry wood was used in making bows.

A tea was made from the inner bark and used as a remedy for diarrhea.

Pueblo Indians dug roots in September, and boiled them with water and a little brown sugar. The tea was taken to relieve inflammation of the stomach. It was also considered good for rheumatism, taken both internally and used in a bath.[2]

currants

wild currant

ALSO CALLED: Wax currant, wild goose-berry, squaw currant, and bear currant.

SCIENTIFIC NAMES: *Ribes inebrians* and *R. cereum.* These two species are very similar and often hybridize completely.

HABITAT and DESCRIPTION: Wild-currant bushes are found from 5,000 to 9,000 feet elevation from South Dakota and Nebraska to Idaho, Colorado, New Mexico, Arizona, and California. The shrubs grow in moist areas of pine forests, they are usually found near other trees and bushes, or at the base of cliffs. The berries, which appear from June to August, are bright to deep red and smooth. The berries are not very juicy.

This is another wild plant that carries an admonition against gluttony. A Hopi writes: "These berries are sweet and tasty and look so nice and red and tempting, but eating too many will make a person sick, so very few (raw berries) can be eaten at one time."[1] We must remember, however, that an Indian out looking for food might make a whole lunch of just this one berry, and when eaten raw in such quantity they are emetic. There is no problem if they are cooked.

The berries were often dried for winter use and much resemble our commercial dried currants, although there is a noticeable seed.

This plant is one of the first to show young leaves in the spring. The Zunis ate the fresh leaves with uncooked mutton fat or with deer fat. The berries were also highly relished.[2]

61

WILD CURRANT PRESERVES
Yield: 1 cup

1¼ cups wild currants
¼ cup water
½ cup honey

Cook water, fruit, and honey together. After the mixture reaches boiling, reduce the heat so it simmers for 20 minutes. Store in the refrigerator. The preserves will become hard when cold but will soften if allowed to warm to room temperature before serving.

CURRANT CORNBREAD
Yield:1 loaf

1 cup toasted cornmeal
½ cup whole wheat flour
½ cup dried ground wild currants
1 tablespoon baking powder
½ teaspoon salt
2 eggs
1 cup milk
¼ cup soft shortening
3 tablespoons honey or ¼ cup sugar

Toast cornmeal lightly by heating in a heavy skillet until brown and fragrant. Combine toasted cornmeal, wheat flour, dried ground wild currants, baking powder, and salt. Add eggs, milk, shortening, and honey or sugar. Beat until smooth. Pour into greased loaf pan and bake at 425° for 20–25 minutes.

OTHER USES: The berries of *R. cereum* were used by the Hopis to relieve stomachache.[3] The wood was used to make arrows.

berries

blossom

elderberry

ALSO CALLED: Blackbead elder.
SCIENTIFIC NAMES: *Sambucus melano-
carpa, S. mexicana, S. neomexicana,*
and *S. coerulea.*
HABITAT and DESCRIPTION: These plants
are large shrubs to small trees. The creamy
white flowers grow in small sprays (botani-
cally called a compound cyme). The edible
fruits are blue-black; *some species that have
red fruits are reported poisonous.* Elderberries
are found throughout the West along washes
and streams and on slopes where there
is adequate moisture. The various species
grow from 1,000 to 9,500 feet.

The western Indians consider the elderberry bush mostly a medicinal plant;
references to the use of berries as food are rare. However, because it is so
widely used for tonics and poultices and because the edible berries grow
so extensively throughout the arid lands, I have included it here.

The sweet, juicy fruits of the elderberry were one of my first experi-
ences of eating wild plants of the Arizona desert. I was delighted to find
a great number of heavily fruiting bushes and trees growing in close prox-
imity to my home. One afternoon, some friends and I gathered a bucket
of berries and crushed them with our hands to make juice. As I was drinking a tall glass of juice, I extolled the great feeling of being able to live off
the land, even in the desert! About a half-hour later that great feeling passed
as I discovered I was one of a small number of people who become nauseous
after consuming raw elderberries.

Since that first experience I have prepared cooked elderberries and elderberry juice in many forms; they have always been tasty and wholesome. *But as a warning, I caution other beginners to confine their elderberry-eating to cooked berries unless they are willing to risk the hazards of experimenting with the raw fruit.*

Depending on the elevation of the tree, it will flower and fruit usually between March and August. Those at lower elevations bloom earlier, and those higher, later.

It is not only the berries of this plant that are eaten. The lovely flowers are also widely used in recipes. When the flowers are used as food, they are called elderblow. If you gather the whole blossom, of course, it cannot develop into a berry. If you have access to a great number of elderberry bushes, this is no problem, but if you want to use the blossoms and the berries from the same plant you can do so by shaking the blossoms into a bucket just before they are ready to fall. Thus, the fruit-producing ovary is not detached and can grow into a juicy little berry.

ELDERBLOW TEA (Three variations on a Cyme)

1. Cover elder blossoms with cold water and let them soak for a day. Strain out the blossoms. Add a squeeze of lemon and sweeten with honey to taste.

2. Since elderberry blossoms and squawberries (see "Squawberry") appear at about the same time, they can both be soaked together overnight, in proportions to your taste. Sweeten with honey.

3. Steep elderberry blossoms and sprigs of wild or domestic mint together in boiling water for 10 minutes. Strain. Flavorsome drunk hot or chilled over ice. A good remedy for a queasy stomach, too.

ELDERBLOW MUFFINS
Yield: 2 dozen muffins

1 cup sifted all-purpose flour
2 teaspoons baking powder
½ teaspoon salt
¼ cup sugar
1½ cups elderblow
2 tablespoons melted butter
1 egg, beaten
½ cup milk
½ cup fruit juice (orange or apple)

Sift together flour, baking powder, salt, and sugar into a bowl. Stir in blossoms. Add remaining ingredients. Stir only until dry ingredients are moistened. Fill greased medium-sized muffin cups ⅔ full. Bake in 400° preheated oven for 20–25 minutes.

ELDERBLOW FRITTERS
Yield: 20–25 fritters

1 cup flour
1 teaspoon baking powder
dash of salt
2 eggs, separated
½ cup orange juice
oil for deep frying
elderblow in clumps on stems or just the flowers

Combine the dry ingredients. Beat the egg yolks slightly and add juice and yolks to the dry ingredients and mix. Beat the egg whites until fluffy. Fold the previously prepared batter into the beaten whites. If you are using the separated flowers, fold them in, too. Heat the oil to 375°. Drop the flower batter into the hot oil by tablespoonfuls and fry and turn until they are golden brown all over. Sprinkle with sugar.

If you are using the bunches of flowers, hold them by the very end and dip them in the batter to coat them. Fry as above.

ELDERBERRY JELLY
Yield: 7 medium jars

3 pounds elderberries
2 lemons
1 box powdered pectin
4½ cups sugar

Put fully ripe berries in a saucepan and crush. Heat gently until juice starts to flow, then cover and simmer for 15 minutes. Place in jelly bag and squeeze out juice. Measure 3 cups of juice. Squeeze lemons and measure ¼ cup juice. Add to elderberry juice.

Mix powdered pectin with juice in saucepan. Over high heat, quickly bring mixture to a hard boil, stirring occasionally. Immediately add sugar. Bring to a full rolling boil and boil hard 1 minute, stirring constantly. Remove from heat. Skim off foam. Pour immediately into sterile jars and seal with paraffin.

Variations: Use half elderberry juice and half strong rhus juice. (See "Squawberry, p. 79.) Follow above recipe.

Use half apple juice and half elderberry juice and follow apple-jelly recipe on the recipe sheet that comes with commercial fruit pectin.

Except for the jelly recipes above, elderberries taste best when they have been dried and then freshened by stewing. Dry them on trays in the sun on their stems. If that very rare rainy day should spoil your plans, they can be dried also in a very slow oven. The following pie recipe includes dried berries.

ELDERBERRY PIE
Yield: 1 9-inch pie

To prepare fruit:
2½ cups dried elderberries
water to cover
⅔ cup sugar

Place berries and water in a saucepan and bring to a boil. Turn down heat and simmer until berries are soft. Remove berries and measure juice. If more than 1 cup, reduce. If less than 1 cup add water or another fruit juice to make 1 cup.

To make pie:
1½ tablespoons quick-cooking tapioca
2 tablespoons water
cooked berries
1 cup fruit juice
⅔–1 cup sugar
½ teaspoon cinnamon
2 tablespoons butter
pastry for 2-crust pie

Combine the tapioca and the water. Combine berries, juice, tapioca and water mixture, sugar and cinnamon. Let stand 10 minutes before filling unbaked pie shell. Dot the filling with the butter. Cover with top crust, crimp edges and cut slits for steam to escape. Bake pie in a 450° oven for 10 minutes. Then reduce the heat to 350° and bake about 40–45 minutes.

ELDERBERRY WINE

There are numerous recipes for homemade wine from elderberries and elderblow. For the most certain degree of success I refer the reader to *Successful Winemaking at Home* by H. E. Bravery (Arc Books). He gives three different recipes for elderberry wine (port style, medium dry, and claret) and a recipe for elderblow wine. His methods are precise but too long to reproduce here. I have not yet made elderberry wine, but it is definitely something I want to do. It seems to me that if you are going to spend time and some money on a project you should opt for the greatest chance for success. So I am going to use one of Bravery's recipes.

OTHER USES: Among the Indians of California the elder was the tree of music and the pithy young stems were used for flutes.[1] The easily hollowed twigs of the elderberry bush also are used for spouts.

Medicinal knowledge of the elderberry and its white flowers is also known in the Far East and has spread through the centuries from the Orient via Europe to the American continent.[2] The Pimas believe the hot tea made from the flowers produces sweating and reduces fever. It is also drunk hot for stomachache, colds, and sore throat.[3] The cold tea from the flowers is reputedly diuretic.

Hot tea made from the bark was used to cure constipation or produce vomiting. And the inner bark fried in mutton tallow was used as an ointment.[4]

A hot poultice of elder leaves is sometimes used on sprains and bruises to relieve the pain and reduce the swelling.[5]

wild grape

ALSO CALLED: Canyon grape.

SCIENTIFIC NAMES: *Vitis arizonica* and *V. californica.*

HABITAT and DESCRIPTION: The several species of wild grapes are found throughout the Southwest from southwestern Oregon down through California, eastward to Utah, Colorado, Arizona, New Mexico, and Texas, at elevations of 2,000 to 7,500 feet. It is common along streams and canyons. The twining vines often climb up trees or dead stumps.

Various groups of Indians used wild grapes whenever they came upon them in their hunt for food. Some grapes were eaten fresh and the remainder were dried for use in flavoring other foods during the winter. The leaves were chewed to allay thirst,[1] a good trick to know when living in an area where water was often hard to find.

Here is a modernized recipe for wild grape.

WILD GRAPE JELLY

Yield: 10 medium jars

About 3 pounds fully ripe grapes
½ cup water
¼ cup strong rhus juice (see "Squawberry," p. 79) or strained lemon juice
7 cups sugar
½ bottle liquid fruit pectin

Stem about three pounds grapes and crush. Add water. Bring juice to a boil, cover, and simmer, 10 minutes. Place in a jelly bag and squeeze out the juice. Let juice set overnight in a bowl. The next morning carefully pour out the juice, leaving behind the tartrate that has settled to the bottom and sides of the bowl.

Measure 3½ cups juice into a very large saucepan. Add the rhus juice or lemon juice and the sugar and mix well. Place over high heat and bring to a boil, stirring constantly. At once, stir in the liquid pectin. Bring to a full rolling boil and boil hard 1 minute, stirring constantly. Remove from heat, skim off foam with metal spoon, and pour quickly into jars. Cover at once with ⅛-inch paraffin.

OTHER USES: Not only southwestern Indians but peoples all over the world, including the ancient Greeks, have used the grape plant medicinally. Grape leaves soaked in water were used as a poultice for wounds and when prepared in various ways were said to cure everything from snake bite to diarrhea to lust.[2]

The Pomo Indians of California used the flexible branches of the wild grape vine in their basketry.

ground cherry

ALSO CALLED: Husk tomato, strawberry tomato, and tomatillo.
SCIENTIFIC NAMES: *Physalis pubescens, P. Fendleri*, and other species.
HABITAT and DESCRIPTION: The ground cherry grows on a leafy, vining plant. Its flowers range from yellow to white to purple, depending on the species. The ground cherry is found in open areas throughout the West on moist to medium dry ground from sea level to 8,500 feet. Various species may be found on plains, mesas, and roadsides along with juniper and pinyon, or along streams in partial shade. The fruit usually appears in the fall.

husk

cherry

Each and every ground cherry comes in its own special container, a lovely little vellumlike "Chinese lantern" that in scientific reality is a five-sided calyx. One of the species is particularly colorful. Its orange-red husks are often grown as a garden ornament and are used in dried-flower arrangements. But the usual field varieties of ground cherry have a green husk, which turns to straw color as fall comes.

Ground cherries can be picked and left in their husks to ripen for several weeks. The ripe berries are in the orange-to-rust range of color.

Most of the arid-lands Indians used the ground cherries and relished them. But the Pimas did not eat them, giving them the "bad name of 'old man's testicles.' "[1] The sweet berries were particularly appealing to Mohave and Yuman children, who ate them fresh when they found them in the fields.[2] Tewa boys cracked the bladdery envelopes with a popping sound by pressing them quickly on their foreheads.[3]

70

The Zunis also relished ground cherries, gathering them in the wild and raising a few in their gardens when they could.

PUEBLO GROUND-CHERRY PRESERVES
Yield: 1–1½ cups

1 cup hulled ground-cherries
1½ cups water
¼ cup honey

Put cherries and water in saucepan; cover and bring to a boil. Reduce heat and simmer for 15–20 minutes or until cherries are soft. Remove cherries with a slotted spoon. Measure juice. If you have more than ½ cup liquid, boil until reduced to ½ cup. If you have less than ½ cup, add water (or perhaps pineapple juice or orange juice) to make ½ cup. Return berries and liquid to saucepan and crush berries with a potato masher. Add honey and cook, stirring often, for 10 minutes. Pour into jelly jar. This is good on toast and very good on ice cream.

ZUNI RELISH
Yield: 1 cup

1 cup ground-cherries
¼ cup chopped onion
1½ teaspoons chili paste
¼ teaspoon ground coriander seeds

Put ground cherries in saucepan and simmer for 15 minutes. Drain. At this point old-time Zuni women got out the stone mortar, but today they probably use an electric blender, so you can too. Combine cooked ground cherries, finely chopped onion, chili paste, and ground coriander in blender. Blend until smooth. Taste and add more chili paste if you like it spicier. Good as a relish with meats.

OTHER USES: No other use of ground cherries seems to have been made.

desert hackberry

ALSO CALLED: Granjeno.
SCIENTIFIC NAMES: *Celtis pallida.*
HABITAT and DESCRIPTION: The desert hackberry is found singly or in dense thickets on foothills and mesas at elevations of 1,500 to 3,500 feet from western Texas to Arizona. The shrub has some spines but feels and looks much spinier than it is because it is so intricately branched. The berries are fleshy orange-red globes with a large seed.

net-leaf hackberry

ALSO CALLED: Sugar berry.
SCIENTIFIC NAMES: *Celtis reticulata* or *C. Douglasii.*
HABITAT and DESCRIPTION: The net-leaf hackberry is a large shrub or tree and is not spiny or intricately branched as is the desert hackberry. It grows near streams or on sides of canyons at elevations of 2,500 to 6,000 feet from Oklahoma and Colorado to northern Mexico. The ripe fruit is orange or brownish-purple and is not nearly as fleshy as the desert hackberry.

Celtis pallida
Desert Hackberry (bush)

Celtis reticulata
Net Leaf Hackberry (tree)

One of the nicest things about desert hackberries, besides their lovely color and sweet taste, is that they ripen in late September. Some of the other desert fruits appear in midsummer, when it is too warm to wander around the desert gathering wild foods. But by the end of September, it becomes pleasant again in desert areas, and it is delightful to get out and hike around. If the hackberry bushes have had sufficient water they will be thick with sweet fruit; if it has been a year of drought, there may be only a few berries scattered here and there on the bush.

The Papagos and the Apaches gathered the fruit and ate it fresh or ground or mashed it into cakes and dried it for winter use. Although the bush did not grow in the area of the lower Colorado Yumans, they would often gather the berries on their wanderings into adjacent territories.

If desert hackberries are put to dry in the sun they shrivel down to currant size and make crunchy, sweet snacks.

72

The fruits of the net-leaf hackberry I have encountered have been so dry and small as to make gathering them impractical. Perhaps in a very wet year it would be different. However, I often pop a few in my mouth while walking down a favorite dry wash of mine. At first, they seem to have no flavor but I hold them in the side of my mouth and soon their lovely, sweet flavor is released.

The following recipe makes a pleasant-tasting though somewhat seedy preserve from the desert hackberry. If you wish a sauce instead of a jam, cook the mixture a shorter time.

HACKBERRY JAM
Yield: 1–1½ cups

1½ cups desert hackberries
1 tablespoon lemon juice
½ cup white sugar
2 tablespoons water

Combine the ingredients in a heavy saucepan. Mash the berries a little with a potato masher to release some of the juice. Bring to a boil, then reduce the heat so the mixture is just simmering. You will have to stir often to keep it from sticking and burning. Cook and stir for ten minutes or until the mixture has boiled down and is quite thick. Remember that the syrup part will thicken as it cools. The purpose of the water is to give the fruit a chance to get a little more cooked before the jam is thickened.

Many people feel that the seediness of the jam made with the above recipe adds extra taste and texture. Certainly it is more authentic. But for those who do not like the seeds, the following recipe for seedless hackberry sauce will give the same delicate flavor, with just a little more work.

HACKBERRY SAUCE
Yield: 1½–2 cups

1 cup desert hackberries
2 cups water
¼ cup orange juice
¾ cup sugar

Combine the hackberries and water and cook gently in a tightly covered sauce pan for 1 hour. Strain the resulting juice and berry pulp into a bowl, working the berries through the sieve with the back of a spoon. This will take some time and concentration.

Discard the seeds and return the pulp and juice to the saucepan. (The pulp will float around in the juice in little globules.) Add the orange juice and bring the mixture to a boil. Add sugar slowly and simmer until the juices have formed a syrup—about 15 minutes. This tasty sauce can be used as an addition to a wide variety of foods—from mushes made of wild seeds to sweet puddings and ice cream.

HACKBERRY BREAD
Yield: 1 9-inch loaf

⅓ cup butter
½ cup brown sugar
2 eggs, well beaten
2 cups whole wheat flour
2 teaspoons baking powder
½ teaspoon salt
½ cup milk
1 cup dried ground desert hackberries
1 tablespoon grated orange rind
½ cup any wild nuts (black walnuts, pinyons, sunflower seeds)

Cream butter with sugar and beat in eggs. Sift together dry ingredients and add to mixture alternately with milk. Add ground hackberries, orange rind, and nuts.

Pour into greased bread pan. Bake at 325° for an hour or more. Cool before slicing.

OTHER USES: No other use, besides food, appears to have been made of the hackberry bush or tree.

manzanita

ALSO CALLED: Bearberry.

SCIENTIFIC NAMES: *Arctostaphylos pringlei*, *A. pungens*, and *A. patula*.

HABITAT and DESCRIPTION: This shrubby evergreen plant grows in dense thickets. The crooked branches are covered with a smooth mahogany-colored bark. Various species grow from 3,500 to 8,500 feet elevation in California, southern Utah, Colorado, Arizona, New Mexico, and Mexico. One species (*A. uva-ursis*) is the Indian tobacco, kinnikinick.

Manzanita is the Spanish word for "little apple," and that is a very good term for the fruit of this plant, as the red berries look just like very tiny apples.

The Indians used the reputedly nutritious berries in a variety of ways. The berries were eaten raw or cooked, made into jelly, and crushed for a beverage. The seeds were sometimes extracted and ground for a mush.

MANZANITA CIDER
Yield: approximately 3 cups

Wash about 4 cups manzanita berries thoroughly to remove all dust and dirt. Place berries in saucepan and cover with water. Simmer very gently (but do not boil) for about 15 minutes or until the berries are soft. Drain the berries, reserving the liquid. Chop, grind, or mash the berries, using a food grinder, blender, or potato masher. The fruit should just be crushed a bit, not reduced to a pulp.

Measure the fruit and put in a bowl. Add as much of the reserved liquid as you have fruit. Let the mixture set for a day and settle. Strain off liquid and refrigerate. Let settle again. Makes a refreshing beverage. This juice can also be mixed with other wild juices for interesting punch combinations.

OTHER USES: A tea made from the berries was used as a lotion to ease the pain of poison oak. A decoction of both the fruits and leaves was used for bronchitis and dropsy, and a tea from the leaves alone was taken for stomach trouble and to reduce fat, and was also used as a bath to aid rheumatism.[1]

Birds, bears, and other animals eat the fruits, and goats and a few other animals sometimes eat the foliage.

rose hips

wild rose

SCIENTIFIC NAMES: *Rosa arizonica,
R. stellata, R. neomexicana*, and other species.
HABITAT and DISTRIBUTION: The different
species are found from 4,000 to 9,000 feet
throughout the West. Some types grow
in pine forests along streams and some in
dry, rocky places. The plants look like typical
rose bushes with prickly stems. The small
berrylike fruits are bright red at maturity.

Roses provide several products that are not only pretty to look at and good to eat, but also contain many vitamins, especially vitamins A, C, and K.

The Indian people lacked our advanced technical knowledge that one rose hip contains 10 mg. of vitamin C or that the seeds are rich in vitamin E, but they did know that they were tasty to munch on and made a good medicine. Hopi children traditionally gathered the fruit while at play in Keams Canyon in northern Arizona.

During World War II, when Britain found itself without enough oranges for its children, vitamin C–rich rose-hip syrup was prepared and given to all the youngsters.

Rose hips are good raw as a snack; rose petals can be served raw in a tossed salad.

77

ROSE-HIP TEA

Collect and dry rose hips. Grind the berries to a powder. Use about 1 teaspoon, more or less to taste, for a cup of boiling water. Let steep for a minute or 2. A little honey makes it even better.

PIONEER ROSE-HIP JAM

Gather rose hips and wash thoroughly. Snip the bud ends with a scissors. While preparing the jam do not use any aluminum or copper utensils, as they destroy the vitamin C.

Weigh the rose hips and use 1 cup of water for each 1 pound of rose hips. Simmer in a covered saucepan for 20 minutes. Rub the cooked pulp through a sieve. Measure the pulp again and add ½ pound of sugar for each pound of pulp. A little ground cloves or cinnamon may also be added. Simmer the mixture until thick—remember that the jam will thicken more on cooling and that you should stop the cooking immediately if you should detect a caramel odor.

Pack in sterilized jars and seal with paraffin. Store in a cool, dry place.

OTHER USES: At Santa Clara pueblo the rose petals were dried, ground fine and mixed with grease to make a salve for sore mouths.[1]

Other groups made a tea from the tender root shoots and drank it for colds. The cooked seeds were taken for muscular pains.

Sometimes the finely ground petals were mixed with sugar, and by means of a paper tube the mixture was blown into the mouth of a person with a sore throat. The water from boiled roses was used as a purgative.[2]

Arrow shafts were occasionally made from the straight branches of older bushes.[3]

squawberry

ALSO CALLED: Lemonade berry and skunk bush.
SCIENTIFIC NAME: *Rhus trilobata*.
HABITAT AND DESCRIPTION: This small bush (rarely over 6 feet tall) grows on slopes and in canyons at altitudes of 2,500 feet to 7,500 feet. It is found throughout the West from Washington to Mexico wherever there is sufficient moisture. The bright red, somewhat flattened berries, which appear in the summer, are usually covered by fine white hairs and are sticky on the outside but not juicy on the inside. The foliage has a distinctive but not unpleasant odor. Each dark green compound leaf has three leaflets.

It is interesting that this plant, which has been of use to desert dwellers over the years in so many ways, is a close relative of common poison ivy, which has caused man so much pain. The two plants look quite different, however.

In the recipes below, which call for meal made from whole ground dried squawberries, grind the berries as finely as possible. Then sift, using a fine sieve. Use only the fine particles that pass through the sieve; the coarser pieces that remain are fine for rhus juice.

RHUS JUICE
Yield: 1 quart

Dry squawberries in the sun. They are very sticky, so this may take some time. Grind berries in a food mill. Combine 1 cup ground squawberries and 4 cups cold water. Let stand for at least 8 hours. Strain off juice. This makes a strong, lemon-y tasting juice. To make a beverage, dilute and sweeten to taste. When camping or hiking, try tossing a handful of berries right off the bush into a pot of water. Let it sit for a while before drinking.

Yavapai leader Grace Mitchell told me that in the old days the Indians sweetened this juice with prepared mescal. The section on mescal tells how to make this natural sweetener, or honey or sugar will do.

This juice can be combined with other wild juices for unusual "wild" punches. It can also add desired tartness when combined with other juices for jelly.

The inventive Apache lady who created the following recipe apparently did so after the arrival of the white man, because it uses white sugar. This unusual, rather tart jam was eaten with bread made of sunflower seeds.[1]

SQUAWBERRY JAM
Yield: 1 small jar or bowl

1 cup water
½ cup sugar
¼ cup ground and sifted squawberries

Combine water, sugar, and squawberries in a small saucepan. Cover pan and bring mixture to a boil. Reduce heat to just simmering and cook for 15 minutes. Uncover pan and turn up heat until mixture comes to a full boil. Stir occasionally. Cook until mixture boils down and becomes somewhat thickened. As the jam becomes thicker you must watch closely and stir almost constantly or it will burn. The cooking time after the lid to the pan is removed is 6 to 8 minutes. Remember that the jam thickens after it cools.

If you are not sure if you have the right consistency, remove the pan from the heat and put a small dab of the jam on a cold saucer. Put the saucer in the freezer compartment of your refrigerator for a few minutes. If it gels and becomes the thickness that you wish, it is done. If not, cook it a minute longer. In any case, as soon as you smell any caramel odor, remove the jam from the heat immediately, before it burns.

Pour into a small jar or bowl.

Apache cooks reportedly made a bread of ground squawberries. I have tried several methods of making bread from this meal. The berries have a very strong flavor, but I find the proportions of berries to flour in the recipe below just right.

SQUAWBERRY BREAD
Yield: 1 9-inch loaf

2 cups whole wheat flour
⅓ cup ground and sifted squawberries

SQUAWBERRY

¼ teaspoon salt
1 teaspoon soda
1 beaten egg
½ cup honey
¼ cup brown sugar
¼ cup vegetable oil
⅔ cup milk

Combine the first four ingredients thoroughly. Combine egg, honey, brown sugar, oil, and milk. Add wet mixture to dry mixture while stirring. Pour into greased bread pan and bake 25 minutes at 425°.

The following recipe came to my attention after fresh-squawberry season had passed, so I have not tested it. It appears in a leaflet written by Elisabeth Hart, a home economist who did work with the Pima Indians in the 1940's. It contains recipes "as told by women who have learned them by word of mouth and by example and their use from those who went before them."[2]

SQUAWBERRY PUDDING

Pick and wash ripe berries. Mash them and put into water. The seeds will sink. Remove berries and cover with water. Boil for 15 or 20 minutes. Using finely ground whole wheat, thicken the berries as much as is desired. Sweeten with sugar. Cool and serve.

(Some people squeeze out the seeds. Some grind the berries to loosen the seeds, then put the berries into water so the seeds will sink.)

OTHER USES: The berries, being strong in tannin, are used by the Hopis as a mordant in dying wool,[3] while the Navajos make a black dye from the roots. A decoction of the roots is also believed to make hair grow.[4]

Among the Hopis the roots are used medicinally in combination with pinyon "for a consumptive."[5]

The buds are used as a deodorant or a perfume.[6]

Several groups use the flexible branches in weaving baskets and cradleboards.

Ceremonially the Hopis used the berries in the preparation of a body paint, and the wood as a fuel in the kivas and for making prayer sticks.[7]

wolf berry

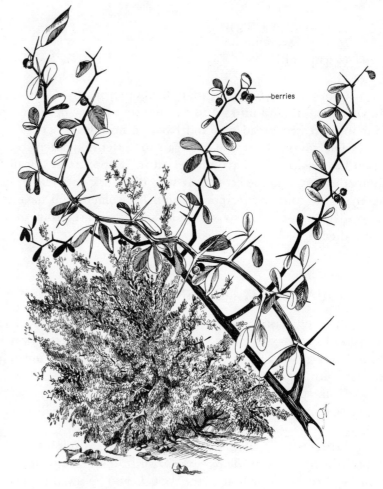

berries

ALSO CALLED: Box thorn, tomatillo, desert thorn, rabbit thorn, and squawberry.

SCIENTIFIC NAMES: *Lycium pallidum*, *L. Fremontii*, and *L. exsertum*.

HABITAT and DESCRIPTION: The various species of the wolf berry are found along washes and on dry slopes of desert and semidesert areas of southern Colorado and Utah, southern California, Arizona, New Mexico, and west Texas at altitudes of 2,500 to 7,000 feet. The many-seeded red to dark red berries of all the species are edible, but some are juicier and fleshier than others. In some cases the berries have a white "bloom." The flowers and fruit generally appear singly on the thorny shrub whose small and smooth-edged leaves usually grow in bundles. The bushes flower early and the berries begin to ripen the middle of May.

The wolf berry plant is sacred to the Zuni bow priesthood. The priests watch the plant, sprinkling meal at its base, awaiting the appearance of the berries. When the berries do appear, the whole plant is sprinkled with cornmeal while the following prayer is intoned: "May the peaches the coming season be as abundant as the berries of the [wolf berry]."[1]

The Zuni word for wolf berry means "water fall down," because the rains cause many berries of the plant to fall off.[2]

It is said that during the famine of 1863 this was the principle fruit eaten by the Hopis, who ground the berries and mixed them with a fine magnesia clay found in the area.[3]

Navajos, Papagos, Yumas, Maricopas, Cocopas, Mohaves, and Zunis were among the other groups who relished the berries fresh off the bush or dried in the sun, when they became much like raisins.

The berries were also mashed in water to form a beverage and were combined in soups and stews. Because the sweetness and juiciness of the berries varies from bush to bush, in an abundant year only the best berries were picked.

WOLF BERRY SYRUP

Boil 2 cups wolf berries in 4 cups water for 1 hour in tightly covered pan. Strain. Reduce remaining liquid by boiling until it has thickened. Combine syrup with water and sugar to taste to make a beverage.

WOLF BERRY SAUCE[4]

1 cup fresh wolf berries
½ cup sugar
½ cup water
flour or cornstarch

Combine berries with sugar and water in heavy saucepan. Cook slowly until berries are tender and cooked. Mash a few with a spoon. Thicken the mixture with a little flour or cornstarch.

OTHER USES: The Hopis used the entire wolf berry plant during their Neman or Home dance.[5]

IV
Foods of Marsh and Mesa

buffalo gourd

ALSO CALLED: Calabazilla, coyote melon, chili coyote, and wild gourd.

SCIENTIFIC NAMES: *Cucurbita foetidissima* and *C. digitata*.

HABITAT and DESCRIPTION: The buffalo gourd grows on a low, trailing vine with numerous stems up to 20 feet long. It is found on roadsides or dry plains and mesas at elevations of 1,000 to 7,000 feet in Missouri and Nebraska to Texas, New Mexico, Arizona, and southern California. The gourd is green with yellow stripes but in the fall it turns straw-colored with paler yellow stripes and is quite conspicuous when the vines dry up.

Cucurbita foetidissima

Cucurbita digitata

The buffalo gourd and the plant on which it grows were used as food, soap, and medicine by the Indians, who collected them when they had begun to dry out in September and October.

Pueblo Indians extracted an oil from the roasted seeds that was used for cooking and for hair oil. They did this by grinding the seeds, pouring boiling water over them, and then straining the vegetable matter from the water after they had soaked for a time. The oil would then rise to the top of the water, where it could be skimmed off. Knochmal reports that seeds of this wild gourd contain 29 percent oil.[1]

Be careful when handling the gourds. The almost invisible fuzz on the fruit can feel like thorns if the fine hairs become lodged in your hand, and they are difficult to find to remove.

86

ROASTED BUFFALO GOURD SEEDS

Dry the gourds until the green is almost gone. This may take 3 or 4 months indoors. Cut the gourds open and separate the seeds from the pulp. Dry the seeds in the sun. Fry seeds in a little oil. Drain and salt. Try combining with other seeds, such as pumpkin, sunflower, and pinyon.

BUFFALO GOURD MUSH

Dry seeds following the method above. Combine with water and cook to a pasty mush. Vary flavor by combining with other seeds. If mush does not appeal to you, use the ground seeds, alone or in combination, dry or in mush form as an addition to breads.

OTHER USES: Several different Indian groups crushed the roots and pith of the buffalo gourd for use in washing clothes, but they were careful to rinse the garments several times to get out the prickly hairs. The Navajos made special trips to gather the gourds, which they used for ceremonial rattles.[2]

There were several medicinal uses for the various parts of the plant. The Pimas pounded the root and boiled it. The extracted juice was put in the ear for ear ache or poured in a hollow tooth to stop the aching. They also claimed this decoction killed maggots in open sores.[3]

The Tewa ground the roots fine, stirred the powder in cold water, and drank it as a laxative.[4] And the western Apaches mashed the stem, leaves, and roots, and soaked them in hot water until the water was soapy. The liquid was then applied to sores on a horse's back.[5]

Some groups believed the top of the plant cured ailments of the head and the roots cured ills of the feet.[6]

Among the Pueblo Indians of the Rio Grande Valley, the buffalo gourd was used as a remedy for rheumatism. The fruits were baked in an oven, split open, and rubbed while still hot on the afflicted parts. An effective purge was made by cleaning out the gourd and filling it with water. After a time, the bitter liquid was drunk by the patient.

cattail

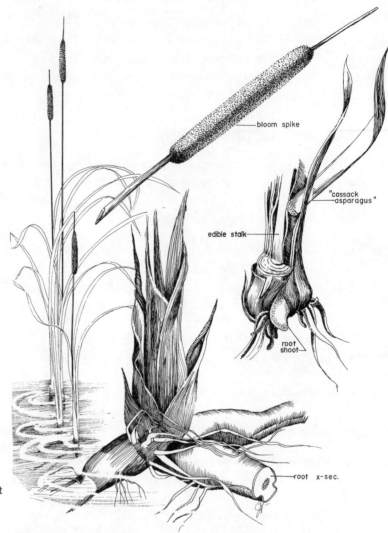

bloom spike

"cossack asparagus"

edible stalk

root shoot

root x-sec.

SCIENTIFIC NAMES: *Typha latifolia* and *T. angustifolia*.

HABITAT and DESCRIPTION: Cattails are found growing in marshes, swamps, and ponds, from low to high elevations all over North America. The leaves are long and flat and the flowers form a dense brown spike on the end of a long, thin stalk.

From its aquatic roots to its flowering tip, the cattail was a veritable food factory for the Indians and pioneers. But despite its inherent value, today it seems to have been somewhat forgotten as a food source.

The roots, the rootshoots, the tips of the new leaves, the inner layers of the stalk, the green bloom spikes, the pollen, and the seeds are all edible. The food value of the root flour is equal to that of corn or rice and the pollen is reported to contain protein, sulfur, and phosphorus.[1]

In the late 1940's, Leland Marsh, a botanist at Syracuse University, did some research on cattails that showed what fantastic yields could be produced if cattails were raised commercially. The rhizomes or roots produced equaled ten times the average yield per acre of potatoes. When flour was produced from the rhizomes, each acre yielded 32 tons, far greater than for wheat, rye, or other grains. Marsh even found a way to ferment the flour to produce ethyl alcohol.

88

Despite this potential, the cattail continues to be ignored except by wild-food enthusiasts.

The modern gatherer who wishes to learn about cattails can begin any time of the year. In the spring there is "Cossack asparagus." To try some, just grasp two or three inner leaves rather low and pull. They will separate from the base of the plant. The tips will be tender, white, and fresh-tasting. Eat them as a snack right there or snip them up for a salad at home.

In summer the green bloom spikes appear. Cut them off with about an inch of stalk attached. Peel off the papery husk if it is still attached. Cover the green spikes with water and boil for 10 minutes. Serve hot with lots of butter, and eat them just as you do corn on the cob. The texture is a bit unusual but the flavor is very good.

Just a little later these green bloom spikes develop a thick coat of bright yellow pollen which can be rubbed or shaken off and used in baking (recipe for muffins below). The traditional Pima method of cooking the pollen was to dig a small hole and build a fire in it. After it was burned down the ashes were pushed aside, cold water sprinkled on the spot, and a layer of powder put on. Pollen and water were layered, covered with leaves and then with ashes, and allowed to bake. The result was reportedly like a sweet biscuit.[2] Yumans made a sweet mush of the pollen, adding a little ground wheat.

To get to the other parts of the plant you might as well resolve yourself to getting a little wet. Wear sneakers and wade in. The rhizomes or roots of the cattail form an interconnecting maze in the muddy ground in which they grow. If you yank up the stalk, the roots may break off and leave the nutritious rhizome in the ground. So slide your hand down the stalk and into the water. Feel in the mud until you get hold of the large root as it runs laterally. Follow it as far as you can and then pull the whole thing out of the water.

Swish the bottom of the plant around in the water until it is fairly clean. The long, brown root looks like thick rope. Coming out of the root and the bottom of the plant you will see little pointed white shoots. These we will call the "root shoots." You also now have the layered stalk and the round bulb at the bottom of the stalk, both of which are edible. Use a big bucket to keep the roots wet as you gather more cattails, for they are difficult to peel if allowed to dry.

BOILED CATTAIL STALK

Cut off the lump from which the roots grow at the bottom of the stalk. Cut off the top of the stalk where it starts getting green. Begin peeling off the layers of the remaining stalk until you reach the very tender inner portion. Discard all but the three or four inner rings. Cut the stalk into half-inch lengths. Boil for ten minutes in lightly salted water. Drain and serve with butter. This is a very good-tasting vegetable. It can also be chopped after being boiled and drained, then sautéed and mixed with cooked rice. Or, include the pieces of raw stalk in a meat stew 10 minutes before the cooking time is up.

ROOT SHOOTS

Cut all the tiny, white shoots off the roots and base of the cattail stalk. Wash well. Slice thinly and use raw in salads, or include with other vegetables in a vegetable stew. The shoots need only a little cooking and are good when served still slightly crunchy. The shoots store well in the refrigerator for a few days if kept in a bowl of water.

CATTAIL CHIPS

For this unusual appetizer you use the lump found at the base of the stalk. Cut off the stalk and the roots and root shoots, leaving a round lump looking sort of like a potato. Wash well and boil for 10 minutes in lightly salted water. Drain well and slice. Fry the slices in hot bacon fat or vegetable oil. Drain and salt. They are best eaten warm.

The potatolike lumps can also be added to a meat stew and cooked until tender. This was a favorite of the Apaches.

MARSH-GOLD SOUP

Cattail pollen adds beautiful color, good taste, nutrition, body, and thickening to any soup. Try a handful or so in a beef or chicken broth. Other wild vegetables (such as cattail root shoots or stalk, wild onions, or wild greens) are delicious and authentic additions.

GOLDEN MUFFINS
Yield: 2 dozen muffins

1 cup cattail pollen
1 cup whole wheat flour
2 teaspoons baking powder
½ teaspoon salt
1 egg beaten
¼ cup oil
⅓ cup honey
1½ cups milk

Sift cattail pollen to remove any debris. Combine dry ingredients. Combine wet ingredients. Quickly mix the two together, keeping stirring to a minimum—10 to 20 seconds is long enough. There will be some lumps. Ignore them. Fill greased or papered muffin cups ⅔ full. Bake for 20 minutes in preheated 400° oven.

Variations: You can add sunflower or black-walnut kernels you have gathered, or substitute any ground seeds you might have for part of the pollen or flour.

There are several methods of making flour from cattail roots—some more complicated than others, some better-tasting than others. Whichever method you use, the roots should be peeled first, before they have had a chance to dry out. The quickest method is to dry the peeled roots, chop them into small pieces, and then grind or pulverize them. When the long fibers are removed, the resultant powder can be used as flour.

The following technique, described by James E. Churchill in *Mother Earth News*, results in a finer product:

Fill a large container with cold clean water. Dump in peeled roots and mash with a potato masher or large stick. When the roots are crushed, reach down and get a bundle and begin tearing them apart. Wring and tear until you have nothing left but what looks like string. Tear until every bit of pulp is gone from the fibers and remove the fibers from the water. Strain the water through a cloth until a lump of pulp and starch is left. Strain with fresh water three times or until the pulp is nice and clean. Dry it and run it through the grinder until all pieces are reduced to powder.[3]

Cattail flour can be substituted for part of the flour in most recipes. Churchill gave the following recipe:

SWAMP BREAD
Yield: 1 serving

½ cup cattail flour
1 large teaspoon baking powder
¼ teaspoon salt

Combine the above dry ingredients and add enough water to form a dough. Form into a flat cake and fry in bacon grease until golden brown, turning to brown both sides.

OTHER USES: Cattail pollen is very sacred to the Apaches. During the last morning of the Sunrise Ceremony (girl's puberty rites), the medicine man mixes cattail pollen and water in a basket. Dipping a soft bunch of grass in the mixture, he walks around the young lady, stopping at each of the four directions, brushing the pollen on her hair. Then he walks around the inner part of the circle made by the crowd and, again using the grass brush, flings the remainder of the pollen over everybody.

The Hopis formerly mixed the brown fuzz with tallow to make a chewing gum.[4] The fluff is also a good insulation for comforters or as a filler for pillows. Indian mothers lined cradleboards with the soft down. The fluff also makes good tinder.

Cattail roots were pounded and mixed with animal fat for a burn salve.[5] And of course the slender leaves have long served as material for weaving rush chair seats as well as calking barrels. Indians wove the rushes into mats and roofs while the Pimas split the flower stalk, dried it, and used it for basketry.[6]

Good torches can be made by cutting off the stalk and dipping the brown spike in coal oil.

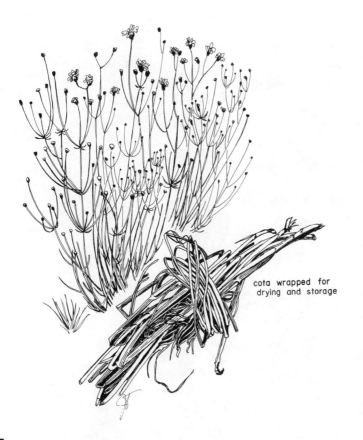

cota wrapped for
drying and storage

cota

ALSO CALLED: Indian tea, Navajo tea, and
greenthread.
SCIENTIFIC NAMES: *Thelesperma gracile*
and *T. megapotaminum.*
HABITAT and DESCRIPTION: Cota is found
in Nebraska and Wyoming to Utah, south
to Texas, New Mexico, and Arizona, and in
Mexico on grassy plains, mesas, and open
woodlands from 4,000 to 7,500 elevation. The
yellow flowerheads of the cota are borne
on long, green stalks and the leaves are very
long and slender. It can be found from May
to October.

This plant has been extensively used for "tea" by practically all Indians of
the Southwest from ancient times to the present.

A Taos Indian family I met one fall had just returned from a round
trip of about 100 miles to gather cota. They prepared it for drying and
storage by wrapping each plant in a circle and tying it in the middle. (See
illustration.)

COTA TEA

Submerge a bundle of fresh or dried cota in a pan of water and boil
for 5 minutes or until tea is the desired strength.

OTHER USES: Apache Daisy Johnson says cota tea with a little sugar
is good for stomachache.

Hopis obtain a reddish brown dye for baskets and textiles from
Thelesperma.[1]

devil's claw

ALSO CALLED: Unicorn plant.

SCIENTIFIC NAME: *Proboscidea parviflora*
(sometimes *Martynia parviflora*).

HABITAT and DESCRIPTION: Devil's claw is
found on plains, mesas and roadsides from
western Texas to southern Nevada, Arizona,
southern California, and northern Mexico
at elevations of 1,000 to 5,000 feet. The plant
is a coarse annual, which grows close to
the ground. All parts are covered with coarse,
sticky hairs. The pink, purple, or yellow
flowers are not abundant but are large and
showy. Leaves are large and sometimes
shallowly lobed.

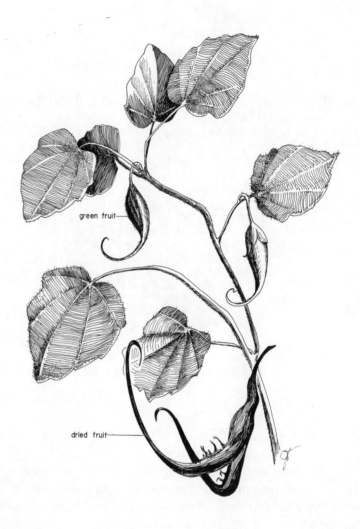

green fruit

dried fruit

Hopis used to say that the long spines of the unicorn fruit drew lightning
and hence rain; therefore, the vine was never weeded out of the fields.[1]

The unicorn fruit is small and hooked on one end when young. When
the fruit matures and dries, the hook splits in two and the pod opens, releas-
ing the seeds. (See illustration.)

In attempting to discover new sources of natural vegetable oils, the
New Mexico A&M College evaluated the seeds of the devil's claw, among
other plants, and found those seeds to contain about 36 percent oil. The oil
resembled cottonseed and sunflower oils and it was concluded that it could
be very satisfactorily substituted for cottonseed oil in the manufacture of
salad oils and shortenings.[2]

94

BOILED UNICORNS

The fruit must be very young to be tender, only an inch or 2 long. Wash under running water, brushing with a vegetable brush to remove as many bristly hairs as possible. Boil in salted water until tender and serve with plenty of butter.

UNICORN SEEDS

Separate seeds from fully ripened pods and dry in the sun. The Pimas cracked the dry seeds between their teeth and ate them like sunflowers. The Papagos ground the dried seeds for mush.

Sometimes "just for fun" the Papagos chewed the fresh seeds and swallowed the juice.[3]

OTHER USES: The outer covering of the pod, when drying, turns black. Both the Pimas and Papagos use this fiber to make the black designs in their baskets.

mormon tea

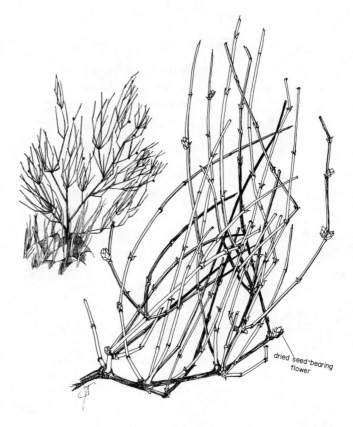

dried seed-bearing flower

ALSO CALLED: Brigham tea, teamster tea, desert tea, and joint fir.

SCIENTIFIC NAME: *Ephedra viridis* and other species.

HABITAT and DESCRIPTION: Mormon tea is found in arid desert grasslands and sage country up to 7,000 feet throughout the western United States. The plants are stiff shrubs with slender jointed branches. The leaves are reduced to scales and occur in pairs or threes. The plants flower in the spring.

Although the strong tea of Ephedra was widely used medicinally as a cure for syphilis, the beverage also makes a pleasant drink for those not so afflicted. Ephedra was used by many southwestern tribes. Because of its activating properties on the mucus membrane, many of the desert tribes chewed ephedra stems to relieve thirst when they were on the move and away from water supplies.

The seeds were roasted and ground for flour or made into a bitter mush.

MORMON TEA

Measure water for the number of cups of tea desired and bring to a boil. Add 1 handful of stems for each cup. (The Navajo first roasted the stems in the campfire.) Remove the pot from the fire and let the tea steep for 20 minutes or more, according to strength desired. Flavor with sugar, lemon, or strawberry jam.[1]

OTHER USES: As a syphilis cure, a large handful of the plant was boiled in a quart of water. The resulting decoction was strained through a cloth and drunk hot three times daily. It was also believed that the tea relieved fever and pain in the kidneys.[2]

The Pimas dried the roots in the sun, powdered them on a flat stone, and sprinkled the powder on all kinds of sores, including those caused by "bad disease."[3] Another group used a decoction of the entire plant as a medicinal beverage to stop bleeding.[4]

The Navajos use *Ephedra viridis* and boil the twigs and leaves with alum to produce a dye of light tan color.[5]

The drug ephedrine, commonly administered as an astringent and as a mild substitute for adrenaline, is obtained from *E. sinica*, a Chinese herb.[6]

When food is scarce, deer and sheep browse on the plant, and quail eat the seeds.[7]

wild onion

ALSO CALLED: Meadow onion.
SCIENTIFIC NAME: *Allium cernuum* and other species.
HABITAT and DESCRIPTION: Wild onions are found distributed in moist ground throughout the Southwest during the spring and summer. The various species range from 1,000 to 10,000 feet elevation and all grow from basal bulbs and have the characteristic onion odor. The leaves may be tubular or flat. While they are usually few in number and slender, some have broad leaves. The single umbrellalike flower ranges in color from orchid, rose pink, and pale pink to white, and is borne on a leafless stalk. Death camas (genus *Zygadenus*) is a poisonous plant that looks similar but lacks the onion odor.

The Tewas and Hopis call this onion the "little prairie onion" to distinguish it from the Spanish onion.

Many Indians ate the wild onion raw in salted water or with corn dumplings or pieces of piki bread. They sometimes roasted the onions in ashes or dried them for use during the winter.

The bulbs of wild onions may be used like regular onions in gravies and stews and cooked in combination with other vegetables. The leaves may be chopped fine and used like chives in salads or sauces.

ROASTED ONIONS

Do not peel onions. Roast on a bed of medium hot coals, turning occasionally. When the onions are tender, puncture the skin to let the steam escape. Scoop out the centers and serve with salt and pepper.

98

BRAISED ONIONS

Parboil the onions in plenty of water and throw the first water away. Return to the kettle with beef or poultry stock to a depth of ½ inch. Simmer the onions, covered, over slow heat for about 20 minutes, letting them absorb the liquid.

CREAM OF WILD ONION SOUP
Yield: 1½ quarts

4 tablespoons butter
1½ cups thinly sliced wild onions
2 tablespoons flour
4 cups milk or part cream
½ teaspoon salt
3 beaten egg yolks
fresh mint (optional)

Sauté the onions in the butter until golden. Stir in flour. Slowly add the milk, stirring while doing so. Add salt. Cover and simmer until the onions are tender. Beat egg yolks in a bowl. Add a little of the hot soup to them, beating with a fork. Add a little more soup and stir. Add this mixture to the soup. Heat but do not boil. Serve topped with a garnish of fresh, chopped wild (or domestic) mint.

OTHER USES: Wild onions were used as an insect repellent by rubbing the whole plant on the skin.

puffball

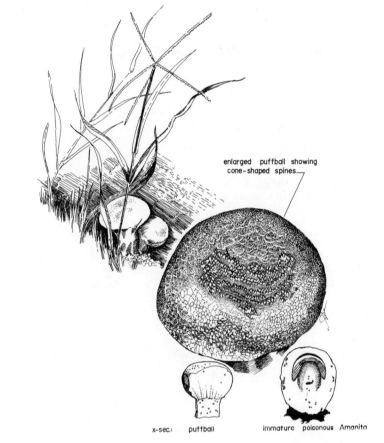

enlarged puffball showing
cone-shaped spines

x-sec: puffball   immature poisonous Amanita

ALSO CALLED: Stomach fungus and
mushroom.
SCIENTIFIC NAMES: *Lycoperdon perlatum*
and *L. pyriforme.*
HABITAT and DESCRIPTION: Puffballs are
found throughout the West growing singly or
in clusters in the rich soil of forests or
occasionally on rotting wood. They appear
only in August or September after the summer
rains have wet usually arid forests. The
above species are round to pear-shaped,
1 to 2 inches in diameter, with or without a
short stem. They vary from white to pale
brown. *L. perlatum* has soft, cone-shaped
spines on the upper surface. These later fall
off, leaving spotlike scars.

Zunis gathered puffballs in great numbers, eating them fresh and drying
them for winter.[1]

*Gatherers of wild foods should exercise a great deal of caution with
mushrooms, as many poisonous types exist.* Puffballs are one of the safest
groups of fungi to gather, however, because there are no poisonous puffballs.

The rule is that any puffball that is white and uniformly the same,
smooth texture inside is good to eat. To check for these two factors, any
puffball that you are contemplating eating must be sliced in half longitudinally, starting from the point of attachment to the ground. If you see an
outline of a stalk, cap, and gills, discard that specimen, for you have the
immature stage of the *Amanita*, a very poisonous fungus. A gelatinous layer
under the skin means that your puffball is not a puffball at all, but the
"egg" stage of a stinkhorn mushroom, generally not considered edible. If
the interior of the puffball is green or yellow, powdery or slimy, it means

100

that it has begun to produce spores and has passed its edible stage. Little holes mean a worm has invaded that fungus. However, if the inside is all creamy white you indeed have an edible puffball. *Check each specimen individually.*

Many folktales are associated with mushroom-eating. Among the Tewas it was believed that a stick must be laid across the top of the kettle containing the cooked mushrooms or the person eating them will be afflicted with a poor memory.[2] The Papagos formerly did not eat mushrooms, saying that they made them old.

Puffballs are good sliced and sautéed in butter and then combined with eggs or other vegetables. They also can be used in the following recipe.

PUFFBALL CASSEROLE
Yield: variable

Clean and slice puffballs longitudinally. In the bottom of a casserole put a layer of sliced puffballs, sprinkle with bread crumbs, and dot with butter. Continue to layer the rest of the puffballs in this manner. Mix together enough beaten eggs and milk in the ratio of 1 egg to 1 cup of milk to cover the mushrooms. Add salt and pepper to taste and pour over the puffballs. Sprinkle the top with grated Parmesan cheese. Bake at 325° until a knife inserted in the custard comes out clean.

OTHER USES: At the pueblos along the Rio Grande Valley, broken eardrums are sometimes treated with powdered puffballs. Spores of the fungus are sprinkled over wounds to help stop the flow of blood.[3]

V
Greens

rocky mountain beeweed

ALSO CALLED: Stink weed, spider flower, and guaco.

SCIENTIFIC NAME: *Cleome serrulata.*

HABITAT and DESCRIPTION: Rocky Mountain beeweed grows in the more arid forests throughout the West at altitudes of 4,500 to 7,000 feet. The stems are tall and branched and there are 3 to 5 leaflets. Attractive purplish-pink or white flowers make it highly visible when in bloom.

Rocky Mountain beeweed was one of the most important native plants to be used by the Pueblo Indians of New Mexico. At Hano pueblo it was considered to be of sufficient economic importance to be named in songs with the three chief cultivated plants: corn, pumpkin, and cotton.

Beeweed is also viewed with some reverence by the old Spanish-Americans of New Mexico who remember stories of the great drought, when their ancestors would have starved but for the seeds of the beeweed, from which they made nourishing if not palatable tortillas.[1]

Other Indian groups that use this green include the Zunis, the Hopis, and the Navajos, who dry the leaves and store them for winter use.

The young plants are gathered up until July, before the flowers appear.

104

Rocky Mountain beeweed is usually prepared by boiling with corn. The Navajos made a stew of it with wild onions, wild celery, and a little tallow or bits of meat. Morsels of bread then were dipped into the stew and eaten.

ZUNI BEEWEED
Yield: 1 quart

1 quart of young beeweed leaves
1 ear young corn
1 tablespoon red chili paste.

Put the washed leaves in a sauce pan and cover with water. Boil for 2 minutes. Drain off the water. Cut the corn off the cob and add to the beeweed leaves. Cover with water and boil until the corn is done. Drain off water. Stir in chili paste.

OTHER USES: The Tewas, Zunis, and San Ildefonso pueblo Indians all make black pottery paint from beeweed. The leaves are boiled in water until the mixture becomes thick and black. Then it is poured on a board to harden. When needed, it is soaked in hot water until soft enough for paint.[2]

Medicinally, the flowers are boiled with a piece of rusty iron and the decoction drunk cold to cure anemia. Sometimes the leaves are crushed and placed on swellings to reduce the inflammation caused by the bites of poisonous insects.[3]

Also, the finely ground plants are mixed with water and the liquid is drunk as a remedy for stomach disorders, or occasionally the fresh plants are wrapped in a cloth and applied to the abdomen.[4]

canaigre

ALSO CALLED: Wild rhubarb and dock.
SCIENTIFIC NAME: *Rumex hymenosepalus.*
HABITAT and DESCRIPTION: Canaigre is
found from Wyoming to Utah, western Texas,
New Mexico, Arizona, and California in
sandy stream beds and fields below 6,000
feet. The long, green leaves appear as early
as February in mild-winter years. Its very
short stems have a reddish tinge. Flowers
and, later, seeds appear on a central stalk that
turns rust red in the fall.

seed spike

leaf

root

Maricopa, Pima, Papago, Navajo, and Hopi Indians were among the groups
that used canaigre either as food or for medicine.

Desert groups, always confronted by a water shortage, often roasted
the succulent leaves in ashes rather than boiling them.

The Maricopas mixed the roasted, ground seeds with water to make
flat cakes, which they baked on ashes,[1] and the Navajos used the seeds in
mush.

BOILED CANAIGRE

Very young canaigre leaves may be boiled in 2 or 3 changes of water
to counteract the bitter taste. They can then be served with butter, or
chopped and fried with onion in bacon grease. They are also good included

106

in soups or mixed with other wild greens. The presence of canaigre leaves among other, more bland greens gives a piquant touch.

The following recipe was collected by home economist Elisabeth Hart on the Pima reservation[2]—I have added the exact measurements. It tastes very much like rhubarb pie.

SIVITCULT PIE
Yield: 1 9-inch pie

4 cups tender canaigre stems
¼ cup flour
1½–2 cups sugar
1 tablespoon butter
1 2-layer pie shell

Collect tender stems and cut into 1-inch lengths. Put in a saucepan with very tightly fitting lid, add just enough water to steam, and cook until stems are tender and juicy. Add sugar, flour, and butter, and cook until sugar is melted and juice has thickened. Pour into pie shell and cover with top crust. Slit top of crust to allow for escaping steam. Bake at 450° for 10 minutes then reduce the heat to 350° and bake about 40–45 minutes longer.

OTHER USES: The roots of canaigre have a high tannin content and have been used by the Indians for tanning leather. Some interest has been directed toward growing rumex commercially for the tannin but none of the attempts has proven successful.

Navajos use the roots to make a brown dye for wool.[3]

Medicinally, the Pimas, Papagos, and Hopis chewed and swallowed the root for bad coughs and colds.[4] Among the Zunis the powdered root was given for sore throat.[5] Sometimes the tuber was boiled, and when cooled the water in which it had been cooked was used as a gargle for sore throat or held in the mouth for sore gums. Both the Pimas and Papagos put the dried, powdered root and a tea made from the root on skin sores.[6]

curly dock

ALSO CALLED: Curly leaf and yellow dock.
SCIENTIFIC NAME: *Rumex crispus.*
HABITAT and DESCRIPTION: Curly dock is found up to 8,000 feet along streams and ditches when there is enough water to sustain it. It was brought from Europe and has spread over temperate North America. The long, narrow green leaves have very short stems and grow from a central cluster. Edges of the leaves are somewhat ruffled—hence the name "curly." A central stalk bears greenish flowers that turn rusty color when the seeds develop. Except for the curly edge of the leaves and a difference in the root, curly dock and canaigre are very similar in appearance.

The leaves of curly dock can be collected when very young and boiled for greens. They should be cooked in several changes of water. One Indian woman said, "The leaves already have vinegar in them and we don't need to add any."[1]

OTHER USES: The Pimas colored the edges of their blankets yellow using dye obtained from the roots of curly dock.[2]

The roots were mashed and used as a poultice on sores and swellings.[3] In the Rio Grande Valley, Indians mashed the leaves, mixed them with salt, and bound the poultice on the forehead to cure a headache.[4]

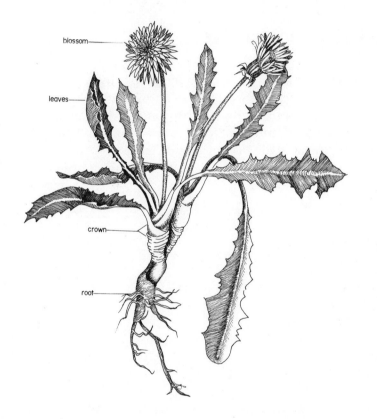

blossom

leaves

crown

root

dandelion

ALSO CALLED. Blowball.

SCIENTIFIC NAME: *Taraxacum officinale.*

HABITAT and DESCRIPTION: The common dandelion is an herb that grows very close to the ground. All the long, toothed leaves emanate from the central core; the flowers are bright yellow and furry and are borne singly on hollow stems. Originally from Europe, the dandelion is now found widely throughout the West in lawns and on roadsides.

Several southwestern Indian groups have adopted the dandelion into the group of plants they use for food and medicines.

The Papagos ate the vitamin-filled young leaves both raw and cooked. Other parts of the plant, including the crown, root, and blossoms are also eaten.

The leaves must be picked when very young, before the plant has flowered, or the characteristic bitter taste will be too strong. When they are very small, the raw leaves make good additions to salads; when the leaves are a little older they are best boiled in salted water for 5 minutes then seasoned with butter and vinegar.

Because dandelion wine naturally comes to mind when talk turns to eating the herb, I have included a recipe for that also.

109

DANDELION CROWNS

Between the green part of the dandelion leaf and the root there is a tender white portion of the plant that we will call the crown. This makes a very good vegetable. Trim off the leaves and roots and wash the crowns well to remove all the grit. Boil in 2 or 3 changes of water until all the bitterness is gone. Cooking time may be as short as 10 minutes in all, as the crowns become tender very quickly. Season with plenty of butter and salt and pepper.

DANDELION ROOT COFFEE
Yield: 1 cup

Scrub dandelion roots with a stiff vegetable brush. Scrape as much of the skin away as possible. Dry the roots in an oven until they are brown and brittle, and then grind them. Boil 1 teaspoon of the ground dried roots with 1 cup of water for 3 minutes. Strain into a cup. This makes a good beverage, especially when laced with sugar and cream.

The remains of the boiled dried roots are quite tasty and can be added to bread or pancake batter.

DANDELION WINE
Yield: 1½ gallons

4 quarts dandelion blossoms (no stems)
4 quarts sugar
4 quarts boiling water
2 lemons
1 orange
1 yeast cake

Cut off every bit of stem from the dandelion blossoms and put blossoms in a stone jar or crock; sprinkle sugar over them and add water. When lukewarm, add the juices and rinds of the lemon and orange and the yeast cake. Stir, cover, and let stand 24 hours. Strain through a fine cloth and let stand 3 days. Strain again and let ferment. Bottle when action ceases. It should remain in bottles for 3 or 4 months. H. E. Bravery[1] also gives a good recipe for dandelion wine.

OTHER USES: The scientific name of dandelion translates into "official remedy." Among the Tewas a fracture was treated with fresh dandelion leaves, which were ground and mixed to a paste with water and spread over the afflicted part, and then whole leaves were bound over the poultice with rags.[2]

110

At Santa Clara the leaves were ground and mixed with dough and applied to a bad bruise.[3]

A tonic for heart trouble was made from the blossoms, which were boiled until the water turned very yellow. A glassful was drunk before breakfast every morning for a month.[4]

A yellow dye for deerskins was also made from the flowers,[5] and a tea made from the green root was considered a good laxative.[6]

lamb's-quarter

seeds

ALSO CALLED: Goosefoot, pigweed, and wild spinach.

SCIENTIFIC NAMES: *Chenopodium album, C. fremontii, C. murale,* and *C. leptophyllum*.

HABITAT and DESCRIPTION: These weeds originated in Europe and have spread widely throughout the West on cultivated lands and other places. The plants are usually branched. The leaves are wider at the bottom than at the top and are sometimes lobed, making the shape of the leaf often resemble the shape of the footprint of a goose. The leaf surfaces are sometimes mealy. The flowers are green and the seeds form heads arranged in spikes.

Lamb's-quarter is one of the mildest-tasting of the wild greens and is a good choice to combine with some of the other strong weeds you might gather to eat.

Used by Indians not only in the Southwest but throughout North America, it has survived as one of the most popular wild foods collected in the United States.

Because of its very moist leaves, lamb's-quarter was often packed around other foods when they were pit-baked. Hopis used it this way when preparing fruits of the yucca.[1] Besides being cooked alone, lamb's-quarter leaves were included in soups and stews. The very young leaves are good raw in salads.

The numerous, easily gathered black seeds also were used, being ground for breads and mushes.

112

COOKED LAMB'S-QUARTER

Choose plants less than 1 foot tall or the new shoots of older plants. These herbs contain much moisture and have no bad flavor to be leached out, so they are best steamed to retain all vitamins. Place leaves and stems in a pot with a tightly fitting lid. Add only a tablespoon or so of water and a sprinkle of salt. Cook over low heat until tender, 5–10 minutes. The cooked greens are good served with butter, vinegar, or a little ground chili.

PIMA VEGETABLE STEW
Yield: 3–4 servings

Soak ¾ cup dried baked cholla buds for 3 hours. Boil for 30 minutes. During the last 5 minutes of the boiling time add 2–3 cups chopped lamb's-quarter. Drain the vegetables when they are tender. In a frying pan sauté very tiny pieces of meat (¼–½ pound) in animal fat. (You may substitute several pieces of chopped bacon and its grease). Add the cooked vegetables and very gently sauté all the ingredients until they are warm and well combined.

LAMB'S-QUARTER SEED BREAD
Yield: 1 small loaf

½ cup ground lamb's-quarter seed
½ cup cornmeal
¾ cup whole wheat flour
2 teaspoons baking powder
sprinkling of salt
½ cup water
1 tablespoon honey
1 tablespoon oil

Combine ground seeds, cornmeal, flour, baking powder and salt in mixing bowl. Add honey and liquids. Form into a ball-shaped loaf on flat baking sheet and bake at 350° for 25 to 35 minutes.

ZUNI STEAMED BREAD
Yield: approximately 10 balls

1⅓ cup boiling water
½ cup cornmeal
½ cup cold water
½ teaspoon salt
½ cup ground lamb's-quarter seed

Bring the 1⅓ cup water to a boil. Combine cornmeal, cold water, and salt, and slowly add to the boiling water. Cover and cook on low heat until the mixture is thick. Remove from heat. Add the ground lamb's-quarter seed. Form mixture into balls the size of golf balls. Place on a rack over boiling water. (The Indians used a group of sticks closely fitted in the bottom of the pot.) Cover the pan and steam the dumplings until done.

NAVAJO GRIDDLE CAKE
Yield: 14 4-inch cakes

¾ cup ground lamb's-quarter seeds
¾ cup whole wheat flour
2 tablespoons sugar
½ teaspoon salt
1 teaspoon baking powder
1 egg
2 cups milk
2 tablespoons bacon drippings

Combine dry ingredients. Beat together egg, milk, and fat and add to dry ingredients. Test the griddle by letting a few drops of cold water fall on it. If the water bounces and sputters, the griddle is ready to use. Bake the pancakes.

OTHER USES: The various species have been boiled and applied as a poultice to reduce swellings and soothe aching teeth. Sometimes a tea made from the leaves was used as a wash to ease rheumatism.[2]

Mentha arvensis
Wild mint

Monarda menthaefolia
Horsemint

wild mint

ALSO CALLED: Field mint.
SCIENTIFIC NAME: *Mentha arvensis*.
HABITAT and DESCRIPTION: Wild mint is found in wet places from 5,000 to 9,000 feet throughout the West. This herb has oval, finely toothed leaves that occur in pairs on opposite sides of a square-stemmed stalk. Pinkish flowers are found at the base of the leaves. The plant has the characteristic odor of mint.

Some southwestern Indians made special trips to the moister areas of nearby mountains just to gather wild mint.

The Hopis and Navajos used mint leaves for flavoring cornmeal mush. Some families at Taos still wrap freshly caught fish in mint leaves before baking their catch in adobe ovens.

Among the best-known uses for mint are tea and jelly. These fragrant leaves can be steeped with boiling water for a refreshing hot beverage or used as a garnish for punches made from other wild juices. Early settlers gathered wild mint to make a good jelly. A modern recipe, with pectin, is given here.

115

WILD MINT JELLY

Yield: 5 medium glasses

> 2 cups mint leaves
> 1 apple
> 2 tablespoons strained lemon juice
> 3½ cups sugar
> green food coloring
> ½ bottle liquid-fruit pectin

Wash mint leaves well and chop. Wash the apple and cut into 8 pieces. Put all in a large saucepan and cover with water. Bring to a boil and cook for 5 minutes, mashing occasionally. Remove from heat and let stand for 10 minutes. Strain liquid and reduce to 1¾ cups.

Add strained lemon juice, sugar, and a few drops of green food coloring. Mix well. Place over high heat, and stirring constantly, bring to a boil. At once, stir in pectin. Bring to a full rolling boil and boil hard 1 minute, stirring constantly. Remove from heat, skim off foam with metal spoon, and pour quickly into glasses. Cover immediately with ⅛-inch hot paraffin.

OTHER USES: Mint tea is a common remedy for nausea, colic, nervous headache, and heartburn.

Mentha arvensis
Wild mint

Monarda menthaefolia
Horsemint

horsemint

OTHER NAMES: Bee balm and wild oregano.
SCIENTIFIC NAME: *Monarda menthaefolia*.
HABITAT and DESCRIPTION: This member of the mint family is found in moist to medium dry soil in Arizona, New Mexico, Colorado, and Wyoming, and north to Alberta, usually at elevations of 5,000 to 8,000 feet. It is a perennial with an upright, usually unbranched stem. The leaves are finely toothed and oval, forming a point on the tip. The purple flowers are in round heads at the top of the stems.

Horsemint has been eaten as a green and used to season other foods. Along the Rio Grande, in the pueblos, it was used to season beans and stews. A good tea may be made from the leaves.

This herb was often dried so it could be used all winter as a spice and as a medicine.

OTHER USES: At San Ildefonso the dried horsemint plant was ground and the powder rubbed over the head as a cure for headache or all over the body as a cure for fever. In the treatment for sore throats the patient was given a tea made from the leaves, and a small pouch stuffed with the leaves was hung around his neck.[1]

117

pigweed

seed spike

mature plant   young plant

ALSO CALLED: Careless-weed, red root, bledo, quelite, and chili puerco.

SCIENTIFIC NAMES: *Amaranthus palmerii* and *A. retroflexus*.

HABITAT and DESCRIPTION: This common, coarse annual is found in disturbed lands such as river bottoms, roadsides, ditch banks, and irrigated fields. Pigweed grows at elevations below 5,500 feet from southern Canada to northern Mexico, usually appearing after the summer rains. When *A. Palmerii* plants have enough water they produce many branches, each branch developing a long spike of tiny greenish flowers at the tip. *A. retroflexus* tends not to branch. The oval leaves are finely toothed and are attached to the stalk by a long stem.

Zunis believed that the seeds of the pigweed were brought from the underworld by the rain priests and scattered by them over the earth.[1]

Among the other Indian groups utilizing the pigweed leaves and seeds were the Hopis, the Pimas, the Papagos, the Havasupais and other Yuman tribes, to whom the pigweed was especially important.

Leaves of pigweed are very mild-tasting, but they must be gathered while young and tender, before the plant blooms.

The tiny seeds are shiny black and can be harvested after the flowers have dried and turned straw color. One Havasupai, Minnie Marshall, demonstrated the traditional Havasupai method of gathering these seeds by stripping the dried flower spike into her palm. She then blew hard on the pile of fluff in her hand. The dried material flew off, leaving a small pile of black seeds.

118

The Yuman tribes that lived along the Colorado River often tied the seedheads together before they ripened so they would be protected. The seeds were harvested after the Indians had finished gathering in their cultivated crops. Seedheads were broken off and taken home, where they were further dried, sometimes under a special shelter built of willow branches. Threshing was done by beating the spikes with sticks on a hardened floor, and winnowing was accomplished by tossing the crushed spikes in a basket so the wind could blow away the chaff. The seeds were parched with coals in a shallow pottery dish and the black seedcoats popped open, revealing the white inside.[2]

The roasted seeds were ground for flour, which was made into bread or mixed with other seeds and grains for mush.

PIGWEED GREENS

Pick young pigweed leaves and wash. Place in a saucepan with a tightly fitting lid and add 1/4 inch of water. Bring to a boil, then turn down heat and steam the greens for 5–10 minutes.

These greens are so mild that they are especially good with just salt and butter. A few drops of vinegar is a good addition. Or the steamed greens may be drained and fried in bacon grease.

The Cocopahs sometimes rolled the cooked greens into a ball and baked them on coals. The mass could then be dried further and stored for winter.

CREAMED GREENS
Yield: 2–3 cups

4 cups pigweed greens and any other wild greens
2 tablespoons butter
2 tablespoons flour
salt
pepper
1 cup milk
1 hard-boiled egg

Choose a pot big enough to accommodate all the greens. Put the type of greens that need the longest cooking or those that need to be cooked in several waters in the pot first and begin cooking. Toward the end of the cooking add the greens that are more tender. While the greens are cooking, make the cream sauce. Melt the butter, take off heat, and stir in flour, salt, and pepper. Cook a minute. Slowly add the milk, stirring constantly. Cook the sauce until thick. Drain the greens and gently mix together with about half as much sauce as you have greens. Put in a serving bowl and garnish with slices of hard-boiled egg.

119

PIGWEED BREAD

Roast pigweed seeds for 45 minutes in a 350° oven, stirring occasionally. Grind and substitute for lamb's-quarter seeds in recipe for Lamb's-Quarter Bread.

OTHER USES: During some years pigweed grows so high and so prolifically that it is cut for hay.

purslane

ALSO CALLED: Pigweed and portulaca.
SCIENTIFIC NAMES: *Trianthema portula-castram* and *P. oleracea.* (Although these two plants are not closely related, they look very alike to someone who is not a trained botanist.)
HABITAT and DESCRIPTION: Both of the above named plants have rather fleshy stems and leaves, and grow very close to the earth. The stems are sometimes pinkish. The leaves tend to have a sheen. Both plants grow in disturbed earth such as gardens. *Trianthema portulacastram* is found in the warmer southern areas, *Portulaca plearacea* in higher, cooler areas of the Southwest. Both usually are found in the early fall after the summer rains.

The slightly sour taste and crunchy texture of purslane make it a good choice for fresh salads. This wild green is so well liked that it is usually available in Mexican markets and occasionally in supermarkets in the southwestern United States.

One woman in the Taos pueblo canned some every year to give to her grown children. "They like that better than regular spinach," she said.

Below is her recipe for preparing purslane.

TAOS PURSLANE
Yield: 2 cups

3 cups fresh purslane
3 strips bacon
flour
water

GREENS   Boil purslane in water until tender. Drain and then take purslane in hands and squeeze all the water out. Lay on paper towels to dry. Chop bacon and brown in frying pan. Remove pan from heat, take out bacon bits, and add flour to remaining grease to make a paste. Add water slowly until you have a nice gravy. Add bacon pieces and purslane, and cook until warmed through.

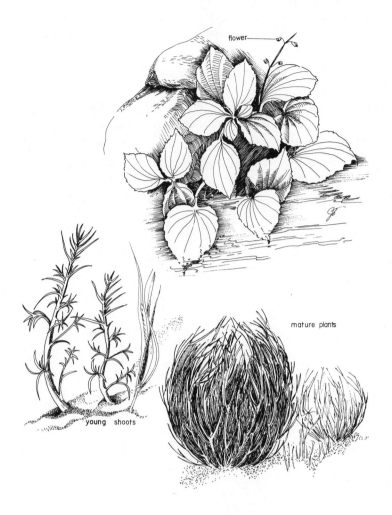

flower

mature plants

young shoots

monkey flower

ALSO CALLED: Wild lettuce.

SCIENTIFIC NAME. *Mimulus guttatus.*

HABITAT and DESCRIPTION: Monkey flower is found widely in the West from Alaska to Mexico and from 500 to 9,500 feet in the very wet ground around brooks and springs. The toothed leaves are round to oval and sometimes slightly hairy; they grow on an erect stem. Flowers are yellow and showy.

tumbleweed

ALSO CALLED: Russian thistle.

SCIENTIFIC NAME: *Salsola kali.*

HABITAT and DESCRIPTION: Tumbleweeds are found widely distributed through North America at elevations of 1,500 to 8,000 feet on roadsides and in fields. As the plants mature they take on a large, round shape, and become very thorny.

The leaves of the monkey flower were gathered by many different groups of Indians and were used for salads and greens. The taste is sometimes slightly bitter and sometimes a little like watercress.

OTHER USES: The raw leaves and stems of the monkey flower were crushed and used as a poultice for rope burns and wounds. The root was used as an astringent.[1]

Large, gray tumbleweeds rolling down deserted roads, being pushed by the wind, are almost a classic signature of the Southwest. The uninitiated would be hard-pressed to understand how anything like *that* could be eaten.

But the Indians were very resourceful in finding good things to eat. And tumbleweed greens are good to eat, but they must be picked and eaten when the first shoots are only 2 to 3 inches tall. If the plants are any larger, they will have developed spines. The new sprouts usually appear after the first summer rains and grow quickly.

Tumbleweed sprouts can be boiled and eaten alone with butter or combined with other vegetables. The very young sprouts can be chopped raw into salads.

The small seeds of the plant also were gathered by the Apaches.

It is difficult to learn to recognize wild greens from a book. The best way to learn them is to take a trip in the field with someone who knows them and can point out the various kinds, but this cannot always be arranged. I found it easier to recognize the young plant if I was familiar with the mature plant. For this reason, even though most wild greens are eaten only when very young, the mature plant is also illustrated here.

VI
Agriculture

Development of domesticated crops was an event of unparalleled importance to the southwestern Indians. The appearance and hybridization of the different crops took place over many centuries and were accompanied by a correspondingly slow but dramatic change in the culture of these peoples.

With the domestication of plants, the Indians finally had a food source over which they had some degree of control. And, of great importance to groups always having to search for their next meal, it meant more food. An increase in the amount of food meant healthier people and lower infant mortality; hence, larger populations.

Because a group that was raising crops was not so nomadic, tending to stay in the area near its fields, feelings of territory were enhanced. Even primitive agriculture produced more food with less work than was required in a system of only hunting and gathering. The extra leisure time allowed the development of arts and crafts.

Corn was the most important domesticated plant to appear in what is now the southwestern United States. It has been recognized as the basic food plant of all the more advanced of the New World cultures.

Very early corn was found in Bat Cave in New Mexico. The tiny cobs excavated in the very earliest layers were only two-thirds of an inch to a little over 1 inch long. Dating by means of radiocarbon analysis places them at about 3600 B.C.[1]

A little later, between 3400 and 2300 B.C., in Mexico in the Tehuacan Valley, hybridized corn was part of an agricultural complex that included gourds, squash, beans, and chili.[2] This agricultural knowledge apparently slowly diffused northward, and a larger type of corn appeared in the Southwest around 500 B.C.[3] By about A.D. 217, the southwestern peoples who were part of the Basketmaker culture had two cultivated crops: corn and squash.[4]

Beans were introduced between A.D. 400 and 700. Their advent was important for two reasons. First, corn is a good source of starch but its protein value varies and it lacks the important amino acid lysine. Beans, on the other hand, are rich in lysine.[5] Second, the cultivation of beans requires a more settled life, for although corn may be planted and then left for long periods of time, beans require almost constant attention.[6]

By the mid-1200's, the Southwest was dotted with little agricultural villages. Then in 1276 a 23-year drought began. The people, accustomed to the fuller life made possible by agriculture, migrated in great numbers to areas north of where the Hopi villages now stand. Here, where a sandstone basin acts as a giant reservoir for underground water, the migrants planted fields and rebuilt their lives.[7]

Wheat did not appear until the Spaniards colonized the area. Father Kino distributed much wheat seed to the Pimas and Papagos in the 1700's and it soon replaced maize as the leading crop among these people.

Each group or tribe of Indians regarded agriculture somewhat differently, depending on its importance to their way of life.

Among the Apaches, agriculture was considered mainly the industry of women. Both men and women took part in planting and the men did help with the work when they were not hunting and raiding, but in general, women were considered better farmers than men. The rituals connected with farming were passed from woman to woman, and women usually selected the seed to be planted. When both men and women set out on gathering expeditions, the old people and very young children who could not travel were left behind to watch the fields.

Crickets figured in the Chiricahua and Mescalero agriculture myths. When the crops had been planted, crickets were placed around the field and enjoined to use their powers to help the crops grow. A live cricket was sometimes put with the seed, and it was believed that if a cricket was killed it would reduce the harvest.[8]

According to one anthropologist, the Cibecue Apaches had more ceremony attached to their agriculture. When the corn was planted, a woman addressed it, saying, "Grow fast; don't bother it, worms; make a good crop." Then, when the tassles formed, a woman who knew the ritual would go to the field and sing, "Make a good ear. Don't hail."[9]

As recently as the beginning of this century residents of some of the more traditional Rio Grande pueblos were allowed to plant only traditional crops—corn, wheat, melons, watermelons, pumpkins, beans, and chili. Any attempt to plant new crops brought punishment. That, however, was a great liberalization of views since the time of the Pueblo Indian revolt against the Spaniards in 1680, when every Spanish importation, including wheat, was burned.[10]

The Indians still maintain some of their old traditions in agriculture, growing varieties of squash, pumpkin, and corn (mostly blue corn), which they relish and which are not found in modern American supermarkets.

The Pimas and Maricopas, living near rivers, developed elaborate irrigation systems for their fields. By means of an aerial survey in 1930, anthropologists were able to detect 124 miles of ancient canals in the Salt River Valley and half that many in the Gila River Valley.[11] Both valleys are in southcentral Arizona.

To the Hopis, agriculture and religion were closely related. Most religious ceremonies were prayers for rainfall and good harvests. Seeds were looked upon as very precious, and only the best were selected for planting.

The old Hopis planted seeds as an indication to the Germ God, who lived in the ground, where and what kind of corn was desired. They were supposed to notify the Germ God where they would plant their different crops, and this diety would then prepare the fully matured crop in advance. Sometimes the farmer would change his mind at the last minute, thus confusing the Germ God. This explained in part why colored corn might appear intermixed with white at times, in reality a natural occurrence of windborne pollination.[12]

Tepary beans - various forms

beans

From aboriginal times to the present, beans of various types have been a very important food in the Southwest. Two distinct types of beans have been used. One type is the pinto, a variety of the common kidney bean. The other is the tepary, which apparently was domesticated by prehistoric Indians from wild plants they found growing in canyons.[1]

The Spaniards found the Papagos growing teparies in 1699. The Indians called the native bean "pawi," and it was one of their staples. In fact, it was so common in their diet that they were called the "Bean People." Apparently when the Spaniards asked the Papagos what they called this food, they answered *"t'pawi"* (It is a bean), and the Spaniards corrupted this to tepary.[2]

Teparies are well adapted to the harsh growing conditions in the Southwest. It has been found that frequently when there is no irrigation water or rain, teparies will produce when regular beans do not. And when ordinary beans are destroyed by insects, teparies often are unaffected.[3]

128

Although the average seed of a tepary bean is a little more than half as large as an average kidney bean, many Indians preferred teparies to beans, as they believed that one square meal of teparies was sufficient for an entire day, whereas other beans had to be eaten twice. Thus, when they traveled somewhere, they took teparies to get the greatest food value from the least weight and bulk.[4]

Teparies had extremely variable forms in the wild state and they retained this characteristic under cultivation. One investigation in 1918 found teparies in round, oval, and flattened shapes variously colored white, speckled yellow, speckled brown, pale flesh, light clay with lavendar speckles, dark brown, greenish yellow, reddish brown, and purplish black.[5]

Among the Zunis, the ritual use of beans of given colors in certain ceremonies has encouraged the cultivation of certain varieties. Each of the six Zuni kivas is represented by a color that is the mark of the compass direction with which the kiva is associated. Initiates into the kiva, usually boys about twelve years old, must bring a bowl of boiled beans of the color appropriate to the kiva they are entering.[6]

To the Yuman tribes living along the lower Colorado, teparies were the second most important crop after corn. Their fields yielded beans of many colors but the white tepary was the favorite and the one most commonly planted, the people believing that it had a better flavor than the others. Sometimes white teparies were ground to make flour. The flour was dampened and kneaded into lumps, then dried and stored.[7]

Beans have been excavated from preceramic archeological sites, and one wonders how those persons, without sturdy cooking pots, were able to prepare the legumes, which generally require prolonged cooking. There is a small bean that the Zunis, within very recent times, have prepared by merely parching and salting. This recent Zuni custom of eating whole parched beans may be a holdover from ancient times.[8]

Beans have been prepared also by boiling, by parching and grinding and mixing with corn or wheat meal, and by cooking with meat. The small green pods were also eaten by many groups.

Teparies are generally not available commercially, so all the recipes given here use the common pinto bean. If teparies are used, they should be soaked 12 hours before cooking. They will swell to twice their original volume and weight.

HOW TO DRY BEANS

Beans can be dried with little loss of nutritional value, and 100 pounds of fresh beans can be reduced to about 10 pounds. Matured beans should be allowed to dry on the vine and then shelled and put on a flat baking pan in a 165°–180° oven for 10–15 minutes. (This, like the Indians' method of parching over live coals, destroys any insect eggs in the beans.)

The beans have dried enough when a few squeezed together in your hand no longer stick together or when you press a single bean between your fingers and no moisture comes to the surface.

The dried beans should then be put in open containers and covered with a clean cloth. Every day for 10 days to two weeks stir the beans. When after several days in a row you notice no evidence of moisture, the beans are ready for permanent storage. Containers should be sealed but not completely air-tight and stored in a warm, dry, well-ventilated room.[9] (This method may be used for drying kernels of corn also.)

BOILED BEANS

Soak beans in cold water to cover overnight. The next day add more water and salt, and boil until beans are tender. A ham bone, a piece of salt pork, or a piece of fatty bacon adds flavor.

REFRIED BEANS

Boiled beans are mashed and extra water added, if necessary, to make a soft paste. Animal fat is heated in a frying pan and the beans are fried in the fat until they are warm and fragrant, and the fat is mixed with the beans. This dish is made by all the southwestern Indian groups, and refried beans are often eaten spread on fry bread or tortillas. (See "Wheat.")

ZUNI BEANS AND MEAT

Pinto or red kidney beans (or teparies) are soaked in water. When they are swollen they are put in a pot with chunks of wild or domestic meat and the whole boiled until the beans are soft. The stew is seasoned with coriander (cilantro), salt, and chili.

MARICOPA BEAN STEW

Equal amounts of dried corn kernels (chicos), dried beans, and dried cholla buds are soaked in water overnight. The next day they are boiled until all the vegetables are tender.

OTHER USES: Hopi priests break ritual fasts by eating tepary beans that have been parched before boiling.[10]

At Santa Clara pueblo, beans were cooked, mashed, and spread on the face to relieve neuralgia.[11]

Some Indians who live along the Rio Grande have borrowed the Mexican custom of treating headache by putting three pinto beans on their temples. If the beans stick to the head it is an indication of illness, but if they fall off it signals that the illness is over and the beans are no longer needed.[12]

130

chili

Mexican Indians, dating back to the Mayas and Aztecs, have used chili with practically everything they eat and relish its peppery ability to lend interest to even the blandest of foods. The chillipiquin, a tiny, round, very hot chili that grows wild on bushes in southern New Mexico and Arizona, was used occasionally by the Papagos, who also traded it to the Pimas. It was not abundant, however, and was considered a real delicacy.

These Indians and others of the southwestern United States, however, apparently did not have the large domesticated chilis until the Spaniards brought them after passing through Mexico.

During the autumn harvest time in all Indian villages, large double strings of the brilliantly red chilis hang from eaves and porches in great abundance. As the chilis dry they turn darker red with some spots becoming dark brown and some spots bleaching to straw color. Green chilis are also used. They are now widely available canned but they can also be dried.

131

Chilis, being high in vitamins A and C, are a nutritious complement to the usual diet of beans and wheat or corn tortillas.

GREEN CHILI

Roast green chilis on charcoal, on a hot griddle, or under the broiler until large brown blisters begin to form and pop on the skin. Turn and roast the other side. Put in a hot, covered pan and steam. Then wrap in a clean dish cloth until cool. The skin should peel off readily. It is the seeds and inside ribs that account for the fiery hotness; remove them and you will have a delicious but still spicy pulp. The fresh, roasted, peeled *chili verde* can now be chopped or mashed for inclusion in beans, stews, or sauces.

To dry the chilis, slit them up one side and take out the seeds. Lay them on a flat pan, salt lightly, and dry in the sun. To prepare dried chilis, soak them in hot water, boil, drain, and mash.

GREEN CHILI SAUCE

Combine mashed fresh or cooked dried green chili with grated onion and pressed garlic juice to taste. Add a dash of oregano and combine. Can be served with meats or beans.

RED CHILI POWDER

Dry red chilis on a string. (This is called a *ristra*.) Break open the dried chilis and remove the seeds, then place on a flat pan in a slow oven until brown. Watch so the chilis do not burn. Cool and put dried chilis through a meat grinder until you have a fine powder.

CHILI CON CARNE

Put hunks of beef or pork stew meat in pot and cover with water. Add salt and pepper. Cook until meat is tender and the liquid is a rich gravy. Remove pieces of meat to a plate and add chili powder to taste to the pot liquor, and bring to a boil. The gravy will thicken somewhat. Return meat to chili sauce and heat through. Add a little garlic juice if desired.

OTHER USES: An aid in reducing the aches of rheumatism is made by soaking a chili that has been split open in warm vinegar for 24 hours. Then a cloth is steeped in the liquid and applied to the ache until it is gone. A wart on the finger can be driven away, it is believed, by wrapping a fresh chili around the finger every day.

Chili seeds and veins are sometimes burned as a fumigant to get rid of bedbugs.[1]

corn

Corn is not simply food to the Indians. To many groups it is the basis of religion and the symbol of fertility and beneficence.

"Seed of Seeds," "Sacred Mother," "Blessed Daughter," and "Giver of Life" are all appellations by which the sacred corn food is addressed. Long tales of how corn came to the Indians and stories of times when the corn maidens visited the ancients were told when the people sat around the fires at night.

One Zuni version of the arrival of corn is that, in the beginning, two witches appeared to the Rain Priest of the North demanding two of his children so the rains would come. He did not want to give up any of his children, but in the end he reluctantly gave the witches one daughter and one son. The witches took the youth and the maiden away and when they slept killed them. Their remains were buried in the earth and the rains fell for four days. On the fifth morning a rumbling noise was heard and the youth appeared from his grave. Again four days of rains came over the fields and mountains and on the fifth morning another rumbling rose from the earth and the girl appeared. The same night the witches planted seeds in the wet earth and the following morning the corn was a foot high and other things had grown and were of good size. By evening all was matured and the people ate of the new food, but everything was hot like pepper. So the raven was called and he came and ate much of the corn. The owl and the coyote also came. By eating the food, they softened and sweetened it so it became palatable to the people. Since that time the fields have to be watched, for the raven takes the corn in the day and the coyote robs the fields at night.[1]

In another Zuni tale, the ancients had to sacrifice seven maidens and each maiden became a different color of corn. These maidens were then resurrected as shiningly beautiful goddesses with great powers and many tales to tell of their intervention into the lives of the earth dwellers.

The Zunis always took great care to secure enough corn in each of the colors for use in ceremonies. Especially necessary were the colors for the cardinal points: yellow corn for north, blue for west, red for south, white for east, all colors for the zenith, and black for the nadir.[2]

The Tewas of New Mexico distinguished seven principal varieties of corn—six colors and a dwarf variety. A myth from Santa Clara tells how long ago the people lived principally on meat until forest fires destroyed the game and the people were starving. The Indians went to a sacred place and danced for many weeks so that the wise elders could have a dream. They finally dreamed that they should make a small hole, place in it pebbles of six colors corresponding to the world regions, and cover the opening with a stone. This they did and the people danced again for several weeks. When they looked in the hole there were six corn plants sprouting in it. From this first planting came the six colored varieties of corn.[3]

In *The Ethnobotony of the Hopi*, Whiting lists twenty varieties of Hopi corn. Multiple legends and beliefs are attached to each color. For example, if a red ear of corn was found in a place where a person had died, it was allowed to remain there for four days, after which it was tucked in the ceiling immediately over the place where it had been lying. If it was still there at the next planting season, he who had the bravest heart took it out and planted it.[4]

Much ceremony was also connected with the planting of corn. Among the Papagos there were songs for every stage of its progress: leaf appearing, stalk growing high, ear forming, and tassel forming.[5] The tribes that lived along the lower Colorado River planted two crops of corn each year, one in February when the trees budded and one in August. As the first seeds were planted members of the corn fraternity would chant, "You are known all over the world, are famous. Come up strong like that and feed everybody so that no one will go hungry." Another person would repeat, "Put me down in your field; bury me. I will be glad to come up."[6] It is much colder to the north, where the Rio Grande Tewas live. They had to wait until April to plant their corn, and always planted with the waxing moon so the seeds could grow with the moon.[7]

134

The ancient Zunis made elaborate scarecrows and strung lines draped with furs, rags, and bones all over their fields. Little hair traps were set up for crows who dared to stop in the corn. If any crows were caught they were taken home and kept in a cage but not fed. When they died the crows were crucified out of doors, head downward, as a warning to their kin to stay away from the fields.[8] But crows were not the worst problem the Zunis had to contend with. It was very dangerous for the ancient Zunis to cultivate their crops unless they went together in large numbers, for their enemies, the Navajos, often took advantage of the time that the Zunis were not within the protection of their fortified pueblo and swooped down on them as they were in their fields.[9]

At harvest time families usually helped one another. At the San Ildefonso pueblo all the inhabitants got together to sweep the plaza before the corn was brought home "because corn is just the same as people and we must have the plaza clean, so that the corn will be glad when we bring it in."[10] The Tewa men would bring in the corn from the fields and the women would shuck it and stack it, selecting the very best ears to be put away for seed corn. Enough seed was put away for two years so that if the next year's corn should fail they would have enough seed left to preserve that pueblo's particular strain of corn. This was very important, as it was believed that the corn of a pueblo was the same as the people. When the women had finished shucking at one household they went to help other relations. Widows, orphans, and needy persons helped at as many huskings as possible, receiving a present of corn at each.[11]

Corn was roasted on coals or in adobe ovens and dried, or it was dried uncooked and then stored on the cob.

The monumental task that remained was the grinding of the corn every time meal was needed for mush or bread. A delightful custom of the Zunis turned this usually monotonous work into play while producing a great deal of high-quality corn flour.

As described by Frank Cushing, an observer who lived among the Zunis around the turn of the century, a number of young Zuni girls would gather early in the morning at the home where the party was to be held. They had a good breakfast. Then about a dozen young men, elaborately dressed and painted, arrived and ceremonially greeted the girls. Eight of the girls then went to the milling stones and knelt to work. An old grandmother began singing a song celebrating the corn goddesses, a man began drumming, and the youths crowded around the drum chanting. Soon the eight girls who were at the metates were grinding in time to the drumming and singing. The girl at one end of the milling line would crush a quantity of toasted kernels, passed the coarse meal to the next girl, who ground it finer and passed it again until, after being passed down the line and growing finer under each stone, the meal came out as fine as pollen. The girls moved their grinding stones up and down in exact time and passed the meal from trough to trough in perfect unison. The girls who were not grinding took ears of corn in each hand and fell into line along the middle of the room, dancing

A White Mountain Apache woman uses the traditional grinding stone. The flat stone on which the grain is ground is called a *metate*. The stone held in her hand is a *mano*. (*Arizona State Museum—University of Arizona, Goodwin Collection.*)

and swaying. All were in motion and rhythm until late in the night.[12]

There were as many ways of preparing corn as the number of different tribes times the number of cooks in each tribe, and seemingly endless variations. Corn was first eaten as a tiny green shoot. When it was time to thin the several sprouts growing from each hill, the plants that were pulled out were boiled and eaten as greens. Many of the recipes call for "green corn." This corresponds to our young sweet corn, or corn at the "milk" stage. Most types of Indian corn become quite hard when mature, so for some dishes the ears must be picked "green."

ZUNI SUCCOTASH
Yield: 1 quart

3 ears of green corn
1 cup partially cooked dried beans
1/2–1 cup tiny pieces of fresh meat
sunflower meal or pinyon meal

Cut the kernels from the ears of corn. Put in a pot with the partially cooked beans and meat and cover with water. Cook until corn, beans, and meat are all tender. Thicken with sunflower meal or pinyon meal, season with salt, and cook until the mixture is almost homogeneous.

ZUNI TRAVELING FOOD

White or yellow corn was boiled with corn-cob ashes until the hulls from the kernels could be removed. It was then dried and toasted, ground, toasted again, ground to a very fine flour, once more toasted, and then sifted.

136

In this manner a whole bushel of corn was reduced to a few quarts of flour. This was combined with very fine, dried meat meal, chili powder, and salt. A single teaspoonful combined with a pint of water made a thickish drink that reputedly was so filling a few sips took care of the most rapacious appetite.[13]

For home use many groups kept a finely ground white cornmeal, which was sweetened and mixed with water for a drink. A powdered preparation of corn meal and sugar can be purchased in food stores selling Mexican foods under the name "pinole de maiz."

APACHE CORN SNACK

Finely ground, parched cornmeal was mixed with ground sunflower seeds, acorns, walnuts, pinyon nuts, or lightly parched pumpkin seeds. This was normally eaten by the pinch. Sometimes it was baked into bread cakes.[14]

The following recipe was demonstrated for me by Ella Tsinnie, a Navajo weaver of some reknown, and also a good cook. This bread got its name because when the batter is wrapped in the corn husks with the pointed end of the husks folded under, it looks like someone kneeling.

NAVAJO KNEEL DOWN BREAD
Yield: 12–14 individual servings

Peel ten ears of fresh green corn, saving the husks. Cover the husks while you work to keep them from drying out. Cut the kernels off the cob with a sharp knife. Put them through a food grinder, using the sharpest knife or the tightest setting (or use a blender or a metate.) Using 1 large piece of corn husk or 2 smaller pieces slightly overlapping, put about 2 or 3 tablespoons of the ground corn toward the top and to one side of the corn husks. Roll the rest of the corn husk around the batter and fold up the bottom, pointed part of the husk. Bake for 1 hour in a 350° oven.

To bake in the traditional way, dig a hole 5 inches deep and big enough to accommodate the amount of bread you wish to bake. Build a large fire in the hole, and after it has burned down, clean out the ashes, put in the bread, cover with a sheet of tin and a layer of damp earth. Build a small fire on top of the dirt.

kneel down bread

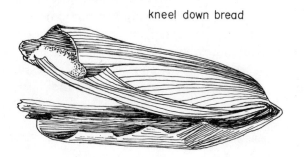

137

The bread will take from 1 hour to 1½ hours to bake. The fire must not be too big or the bread will burn. (With the addition of a thin strip of green chili and a small stick of white cheese wrapped in the center of the dough, these breads become green corn tamales.)

APACHE ASH BREAD
Yield: 3 small leaves

Strain 1 cup of hot water mixed with juniper ashes through a grass brush (or strainer) into 2 cups of cornmeal, and mix to form a firm dough. Add a little salt and knead. Form into 3 flat cakes. Place the loaves on a bed of ashes and cover with ashes and small coals. After about 1 hour, uncover the bread, and if it is not completely done turn the loaves over and recover with ashes and coals. When the bread is done it should be placed in a basket and water poured over it to remove the ashes.[15]

The following cake is the traditional food for the Navajo girl's puberty ceremony. The usual cake is 8 inches deep and 5 feet across and takes all night to bake. The girl and her guests sing and dance all night and everybody has a piece of the corncake in the morning. The first piece cut is from the center and it is divided into 4 parts for the main singer and those who helped him.[16] This recipe has been scaled down for a smaller pit and will bake in a few hours.

KINAALDA CAKE
Yield: 1 cake 18 inches in diameter

½ pail boiling water
6 handfuls white corn flour
1 cup sprouted wheat or sprouted oat flour*
3 cups cold water
1 cup brown sugar
½ to 1 cup raisins
corn husks

First, dig a hole 6 inches deep and 18 inches in diameter. Build a fire in the pit and burn it for 4 hours. Put the half pail of water on the fire (or stove) and bring to a boil. Mix corn flour and sprouted wheat with cold water to make a paste. Add more cold water if necessary. When water in pail is boiling, stir in the corn flour and the sprouted wheat or oat-flour mixture. Stir until all is mixed well. Remove pail from fire and cool. Reach into the mush with your hands and squeeze all the lumps until the mush is very smooth. This may take some time. Add the sugar and raisins. Dip the

* Grains of wheat or oats can be sprouted on a damp, clean dish towel (after being soaked for 24 hours), then dried and ground. Or this meal may be found packaged in some foodstores in the Southwest under the name panocha flour.

138

corn husks in water to soften them. Rake the ashes out of the pit and line the bottom and sides of the pit with the softened corn husks, weaving them together. Pour the mush into the pit and sprinkle cornmeal on the surface from east to west, south to north, and around sunwise to make the cake holy. Cover the batter with another layer of corn husks and a layer of wet cardboard. Then add a layer of dirt and the hot coals. Build a small fire on top of the cake and let it bake for 3–4 hours.

The cook at the Tuba City jail contributed the following recipe. The inmates were unfortunate to be in jail but lucky to have Loretta Blatchford cooking for them.

NAVAJO ROASTED CORN

Build a fire in an adobe oven and keep it going all day long. Clean out all the ashes and put in the ears of corn, husk and all, and ½ bucket of water. Seal the oven with a rock and mud. Seal with mud any places where steam is escaping from the oven. Let the corn bake all night. In the morning scrape the kernels off the ear and dry in the sun.

The Apaches roasted corn in underground ovens, but no one who had been struck by lightning was allowed to help at this task, for it was believed that the corn would not cook under this condition.[17]

An Apache woman, Daisy Johnson, began her recipe for cooking corn this way: "Take three pickup loads of corn. . . ." You might not want to make *that* much, but this is a good method of cooking corn for a crowd.

APACHE ROASTED CORN

Dig a very deep, very big hole. Put a great deal of fire wood in the hole, put many big rocks on top of the wood and more wood on top of the rocks. This was usually done by 6 or 7 women all working together. The fire is traditionally lit about three o'clock in the afternoon. When the fire dies down, cover the rocks and coals with corn husks and corn leaves. Fill the pit with ears of corn still in the husk. Place a stovepipe in the middle of the pit, reaching partly into the piled corn. Pile corn stalks thickly on top of the ears and cover with large sheets of canvas, allowing the stovepipe to protrude. Pile dirt on top of the canvas. For each 100 pounds of corn in the pit, pour 5 gallons of water down the pipe and plug up the hole. The corn will be ready the next morning. (The Apache way is to uncover it at sunrise.) If the corn is to be stored for winter it must be cut from the cob and put to dry in the sun.

CHICOS

In New Mexico, dried kernels of corn are called chicos. To cook dried corn, cover with water and soak kernels until swollen. Add enough water to cover swollen kernels, put in a little sugar, and bring to a boil. Turn heat down and cook until almost tender. Add salt, pepper (or chili), and butter to taste and finish cooking. The presoaked chicos also may be added to meat stew.

The outer skin of some of the early Indian corn was very tough, and plain boiling was often not enough to soften it. The following method of making hominy evolved as a way to get rid of the tough skin. In legends hominy was often included in the basket of food carried by a young Pueblo maiden to the man of her choice.

HOMINY
Yield: 2 quarts corn

1 quart wood ashes or corncob ashes
or
4 heaping tablespoons powdered lime
4 quarts water
2 quarts fresh corn kernels

Using a granite or enamel pan (not metal), add the ashes to the water and boil for 30 minutes. Stir until the mixture stops bubbling, and strain. If using powdered lime, simply dissolve the lime in the water. Add the 2 quarts of corn kernels or as much corn as the water will cover. Cook until the hulls loosen from the kernels. Remove the corn from the heat, drain the liquid, and wash until all the lime taste is gone.

When the Spaniards arrived in the New World they found the Mexican Indians eating little flat cakes made of cornmeal. They called this new food "tortillas" (from the Spanish word *torta*, meaning tart or cake), and possibly introduced this method of cooking corn to the inhabitants of what is now the southwestern United States.

TORTILLAS

Grind hominy into a paste on a metate or run through a food grinder several times. Then shape the dough into balls the size of a large walnut and flatten very thin by slapping between your palms or rolling with a rolling pin, or by using a Mexican tortilla press. If the tortilla looses its circular shape after being rolled out, you can make it round by placing a saucer on the rolled-out dough and cutting around it with a knife. Bake the tortilla quickly on a hot ungreased griddle.

(Modern note: If using a rolling pin or a tortilla press, put a piece of waxed paper both under and on top of the dough. Peel the paper from one side of the tortilla and put that side down on the griddle. When the

140

tortilla has baked just a few seconds, the top piece of waxed paper will loosen and can then be peeled off easily.)

The most popular corn among the Navajos, the Hopis, and the Indians of the pueblos is blue corn. The color at first may seem a little strange to the uninitiated, but the flavor is superior to that of white or yellow corn. Although blue cornmeal may be substituted for white or yellow cornmeal in any recipe, the dishes below are traditionally made with blue corn. Ashes of various plants, including juniper and salt bush (*Atriplex canesens*), are used so the corn will remain blue and not turn green or gray. It also serves as baking powder.

A Navajo teacher took time out from her busy life to give me the first three recipes below. Navajo blue bread is excellent.

Navajo and Hopi women use a cooking utensil that is made of many thin sticks of sage or other grasses tied in the middle. This brush is used to stir mush and as a strainer to catch pieces of ash as the water in which the ash has been soaking is poured into the cornmeal.

NAVAJO BLUE BREAD
Yield: 16–20 3-inch cakes

2 tablespoons juniper ashes
1 cup boiling water
2 cups fine blue cornmeal
sprinkle of salt, optional
pinch of sugar, optional

Put the juniper ashes in a little bowl and add the boiling water and stir. Strain the hot water into another bowl into which you have measured the cornmeal and salt and sugar, if desired. Stir and mix until it is like biscuit dough, adding more water if necessary. Let the dough cool. Roll 2 tablespoons of the dough into a ball and flatten to a 3-inch cake. Lightly grease a griddle and bake the cakes for about 10 minutes on each side. Navajos eat the cakes dipped quickly in salt water.

BLUE CORN PANCAKES
Yield: about 20 4-inch cakes

3 cups blue cornmeal
1 cup white flour
goat milk

Combine the cornmeal, flour, and enough goat milk to make a thin batter. Cook the pancakes on a hot, flat stone or a griddle greased with mutton fat.

141

BLUE CORN MUSH WITH ONION GRAVY
Yield: 4–6 servings

small pieces of mutton fat
2 cups goat milk
1 cup blue cornmeal
1 teaspoon juniper ashes
3 tablespoons water
2 tablespoons sliced wild or domestic onions
white cornmeal

Brown small pieces of mutton fat in a frying pan. Remove the remaining solid material, leaving grease in pan. Combine goat milk, blue cornmeal, and fried mutton pieces. Stir ashes with water and strain 2 tablespoons of water into the mush mixture. Boil together for 15 to 20 minutes.

Meanwhile, fry the sliced wild or domestic onions in the leftover mutton grease, and add white cornmeal, salt, and a little water to make a gravy. Serve over the mush.

The next two recipes were given to me by Bessie Kewenvoyouma of Moenkopi, a Hopi village entirely surrounded by the Navajo reservation. The blue corn dumplings turn very blue and are very good. They form their own gravy.

HOPI BLUE CORN DUMPLINGS
Yield: 8 dumplings

1 tablespoon juniper ashes
½ cup boiling water
1 cup fine blue cornmeal
sprinkle of salt

Mix ashes with boiling water and strain water into a bowl with the cornmeal and the salt. Mix, adding more cornmeal if necessary to form a stiff batter. Roll batter into balls the size of large marbles. Put in a frying pan and pour boiling water over them. Boil for 15 to 20 minutes. Eat with meat stew or boiled eggs.

BLUE CORN CAKE (leaves room for improvization)
Yield: 1 9-inch round cake

a large handful of blue cornmeal (about 2 cups)
2 to 4 beaten eggs
2 teaspoons baking powder
dash of salt

142

Mix the ingredients together. If few eggs are used add enough water to make the mixture the consistency of cake batter. Bake in a greased frying pan in a very hot oven until done. When eating this cake the Hopis dip each piece in water mixed with a little chili powder.

(Modern note: I like this corn cake better if 2 tablespoons of sugar are added to the batter.)

PIKI

Piki is the original Indian bread. It is of Hopi origin but it is made in all the pueblos—the Zunis call it hewe, the Tewa tribes along the Rio Grande call it mowa, and at San Ildefonso it is called bowahejahui, which translates "put it on, take it off." Yumans also made a type of wafer bread.

Essentially it is all the same thing: blue cornmeal batter that is baked in large tissue-paper thin sheets, which are rolled up like a newspaper. Occasionally white or pink piki is made for special dances.

Piki-making is an art and a ritual. It takes years of practice to become a good piki baker. Years ago a young woman was required to demonstrate that she had mastered the art of piki-baking before she was considered a suitable bride. But today few young women have the patience to spend long, hot hours in front of the piki stone, and the number of those who excel in this art is dwindling.

Besides being a household staple in Hopi villages, piki is also a ritual food that is fed to the kachinas during dances.

My own experience with piki-baking was one of the most interesting days I spent while doing research for this book. My patient teachers were Nancy Howard and her sister Bessie Kewenvoyouma, both Hopi ladies living at Moenkopi.

The first requirement for piki-baking is a piki stone, which is large, flat, and dark gray. The size is approximately 4 inches thick, 28 inches long, and 18 inches wide. The piki stone is raised on four stone legs, one at each corner. Formerly, women would go together to gather and prepare the stones —a very precise, solemn ritual. But now several women in the village do the gathering and preparing and sell the stones to the other women for about $25 apiece.

The next thing that is required for making piki is to gather and chop enough cedar (or juniper) wood to build a fire under the piki stone and keep it going during the duration of the cooking. After the fire is lit, crushed watermelon seeds are scattered on top of the stone.

The piki batter is prepared somewhat differently among each group of Indians, but all require ashes, blue cornmeal, and water. Nancy Howard made her piki batter by mixing some ashes of the salt bush (ashes of bean vines, juniper trees, and corn cobs are variously used) with about a cup of cold water. Then she took a large bowl and filled it about a third to half full of very fine, blue cornmeal, as fine as flour. To this she added a quantity of boiling water and mixed the water and cornmeal together. The ash water was strained into the mixture through a bundle of fine grasses. She continued mixing the batter with a wooden stick, finishing the kneading by hand. More and more cold water was added until the mixture resembled a very smooth pancake batter.

By this time the watermelon seeds were well-browned, indicating that the stone was hot enough for baking. Nancy Howard knelt in front of the stone and lit a splinter of wood, which she used to scrape the seeds together in a pile in the middle of the stone. She then lit the seeds with the splinter and when they were black she rubbed them all over the top of the stone with a clean cloth to grease the surface. A thin layer of batter was spread on the stone and immediately wiped off and all was ready for baking.

No matter how difficult a task, if it is performed by an extremely skilled person it seems effortless. Dipping her fingers into the batter, Nancy Howard ran her hand from left to right lightly over the hot stone, leaving behind a swipe of batter that looked something like the first time you run a wet sponge over a dirty window. Two more swipes made a rectangle about 14 inches by 8 inches. A few more light touches patched up any places that had no batter, and a stroke on either side of the rectangle sealed the edges. A few seconds of baking and the transparently thin sheet began to brown and separate at the edges. With one smooth movement, she lifted the sheet from the stone and placed it on a small table beside her. Quickly she smeared another sheet on the stone. During the few seconds it was baking, she laid the first sheet she had baked on top of the bread presently baking, let it warm for just a second, then folded it in thirds, rolled it into a neat roll, about the size of an ear of corn, and placed it on a plate beside her. Keeping up this rhythm she soon had a plate piled high with neat, fragrant, blue-gray rolls of piki.

By this time several other women had gathered to chat, a common occurrence during piki making, and children, attracted by the delicious smell, raced through the little room, grabbing rolls of the relished piki as they ran by.

Then Nancy Howard motioned to me—it was my turn at the piki stone. I begged off—this was to be a demonstration, not a teaching session. "Go on, try," they all urged. Overruled, I knelt in front of the piki stone. Now, it is one thing to marvel at the expertise of a piki-baker and speak of how the art must be perpetuated, but quite another to contemplate burning your own hand on the 700° stone. The principle is to get and keep just enough batter on your fingers so they will glide over the face of the stone with just the very thinnest layer of batter separating them from the burning surface. I tried, but every time I got my hand on the stone I was afraid of burning myself and either left a glob of batter on the stone or succeeded in flinging it all over everybody sitting on my right.

When, after several attempts, I finally made a sheet that was very small and very thick but vaguely resembled piki, I set it aside on the table. Just then a little Hopi boy about six years old wandered through the room for the first time. He was going to snatch a roll of piki when he noticed my poor attempt. He picked it up and got the most comical look on his face,

144

as if to say, "What in heaven's name is THIS?" The old women all laughed uproariously. As a matter of fact, the jeers and guffaws continued the entire time I was seated in front of the stone. Finally, in exhaustion and embarrassment, I insisted that my teacher, Nancy Howard, take over again. Later she explained to me that this was the way Hopi maidens traditionally learned to bake piki. Apparently, being Hopi did not make their first attempts any better than mine and the taunts and jeers of the older women made them more determined to learn the art and prove themselves real women.

Fermented corn liquor was made by several groups and was called both tiswino and tulbai. The old Apache recipe calls for soaking corn and sprouting it until the seedlings are ½-inch long, grinding the sprouts, and boiling them. The resulting mash was then sweetened with mesquite flour or saguaro syrup and allowed to ferment in a brewing jar that was never washed so as to retain the organisms for fermentation. The finished product had to be drunk within a few hours after it was prepared or the alcohol became acetic acid, giving the beverage a sour taste.

The following recipe for corn liquor[18] is the method by which some New Mexican Indians made tiswino, spelled "tesguino" there. I have not tried this recipe but it sounds good.

TESGUINO
Yield: 4 gallons

10 pounds dried corn
4 gallons water
8 piloncillo cones (pyramid-shaped brown sugar)
6 cinnamon sticks
peel from 3 oranges

Roast corn to a pale brown in the oven. Grind coarsely and put in a crock. Add the water, piloncillo, cinnamon, and orange peel. If the weather is warm let this remain for 4 or 5 days. If it is cool the fermentation will take somewhat longer. Strain through a cheesecloth.

Sue and Frank Lobo, anthropology graduate students at the University of Arizona, developed the following shortcut for tulbai, after their attempts with the original method failed. Frank is a mission Indian from San Juan Capistrano, and both he and his wife are interested in Indian foods.

SHORTCUT TULBAI
Yield: 3½ gallons

7½ pounds of raw masa (this can be purchased at a Mexican
    delicatessen or made by grinding hominy into a mush)
10½ quarts warm water
1 pound, 10 ounces piloncillo (brown sugar cones)

145

Heat some of the water to boiling and pour over the piloncillo cones to dissolve them. Add the rest of the water and the masa and stir. Place in a large clay olla (jar) and cover with cheese cloth. This will be ready in about 4 days.

The tulbai has to be consumed on the fourth day, as the quality begins to deteriorate after that.

OTHER USES: Corn and the mention of corn is a conspicuous feature in the ceremonies of many southwestern Indians.

Among the Papagos, almost all ceremonial orations, even those for war, mention in some way the ideal good fortune—a good corn crop. When salt was procured from the Gulf of California—an important ceremonial occasion—the salt was called "corn" and if a great deal was brought home it helped toward a good corn crop. All the Papago ceremonial cornmeal came from "flatheaded corn"—twin ears which were flattened where they joined.[19]

During religious dances at Hopi or Tewa pueblos, sacred cornmeal is always sprinkled on the dance floor and on the dancers and ears of corn figure prominently in the ceremonies. Zunis traditionally forced a delicate mush of white meal down the throat of a dying rain priest so that he would not fail to have sufficient nourishment during his four-days' journey to reach the underworld.[20] And at Hano, when a rabbit was given to a woman she laid it on the floor and dropped meal on it "to feed it."[21] Also at Hano, a corncob tied to the kiva ladder, and "swinging in the smoke" means that one of the men who attend that fraternal society has not yet brought his contribution of firewood for the ceremony in progress and that symbolically his comrades are reduced to burning cobs for fuel because of his negligence.[22]

Corn was also used medicinally. At Santa Clara a child with swollen glands was told to roll an ear of corn back and forth on a warm hearth with his foot and the swelling was expected to subside in two or three days. At San Ildefonso corn pollen was recommended for palpitations of the heart.[23] Many Indian mothers rubbed fine cornmeal on children's rashes.

Blue cornmeal has always been considered more nutritious than other colors. Among Navajos it was thought an especially appropriate food to include in the diet of pregnant women from the onset of labor to the end of lactation.[24]

Corn smut, which sometimes infests the ears, was not regarded as all bad. Apache girls rubbed it on their faces to bleach their skin.[25]

At Zuni, a pinch of smut was put into a small quantity of water and taken at intervals by women giving birth to hasten the arrival of the baby by increasing the severity of labor. It was also given to stop hemorrhage after giving birth.[26] This same decoction was drunk at San Ildefonso as a cure for diarrhea and at Santa Clara it was used by women as a remedy for irregular menstruation.[27]

146

melon

Watermelons earned the nickname "horse pumpkin" among the Hopis because they said a fresh watermelon smelled like a sweating horse.[1] Because the Indians did not have horses before the arrival of the Spaniards, this nickname supports the belief that watermelons also were not known before the Spaniards' arrival.

However, this fruit must have been adopted right away because Father Kino wrote in October 1700 that the Yumans he had visited had watermelons, and the Anza expeditions in the fall of 1775 recorded that the Yumans offered them an estimated three thousand watermelons.[2]

Although not an aboriginal crop, watermelons have been mentioned in ritual songs at Santa Clara as one of the principal crops.[3]

All of the tribes seemed to grow at least two different types of watermelons—one with pink seeds and yellow flesh and one with black seeds and yellow flesh.

Watermelons were aways highly relished. The early varieties had very good keeping qualities and could be packed in weeds and sand or hung in yucca fiber slings for storage until as late as February, in some cases. This was an important source of fresh food during the winter. Among the Pueblo groups, watermelons were often given as gifts on ceremonial occasions.

The fruit was usually eaten fresh, but sometimes the pulp was squeezed between the hands and the juice strained through a cloth and boiled to a syrup.[4]

Mohaves, Yumas, Cocopas, and Maricopas all ate watermelon seeds whole after they had been parched. Sometimes the oily meal was mixed with mesquite flour.[5]

Musk melons were also grown by many of the groups, although they were held in less esteem than watermelons. Many were eaten fresh, but because they could not be stored whole like watermelons, they were dried for winter use. The melons were peeled, and after the insides were scooped out, they were hung on a branch to dry for a day. This made the flesh more pliable so that it could be torn into a spiral strip. The strips were bundled and hung in the storerooms.

During the winter the dried melons were chopped fine, stewed with sugar or honey, and used as filling for pies.[6]

Muskmelon seeds were sometimes parched and eaten. The Mohaves and Yumans boiled the seeds and then ground them on the metate to make a thick gruel, which they regarded as a delicacy.[7]

OTHER USES: Crushed watermelon seeds are used to grease the stone on which the traditional Indian bread, piki, is baked.

pumpkin

Every year, about the time the pumpkin runners were about to spread, Apache families held a little ceremony designed to ensure a good pumpkin harvest. A young boy was sent out to pick a large supply of blue juniper berries. When he returned he was blindfolded and sent into the pumpkin patch, where he threw the berries in all directions, asking for that many pumpkins. After this, newlyweds or pregnant women could not enter the field of growing pumpkins, for fear the fruit would shrivel and die on the vine.[1]

Pumpkins were cultivated at all the Rio Grande pueblos at the time of the arrival of the Spaniards and are still a fairly important crop. Pumpkins also were grown by the Hopi, Navajo, Havasupai, Papago, Pima, and Yuman tribes, among others.

Several varieties of pumpkin were grown, including the common orange field pumpkin and the cushaw type. A kind of pumpkin that remains

149

green on the outside at maturity is still cultivated and referred to as "Indian pumpkin."

Pumpkins were cut into pieces and boiled or baked whole in the adobe bread ovens or directly on coals.

All of the groups dried quantities of pumpkin for winter use. The pumpkin was usually peeled and then sliced into rings or cut into a continuous spiral. The dried bundles were tied with yucca fiber and stored in jars or hung up in storerooms. The Indian pumpkins were thought to be the best for drying. "When they are boiled up they taste just like fresh," I was told at Taos.

Today we usually prepare pumpkin with some sweetening. The Maricopas liked their pumpkin sweet too, and put a layer of mesquite beans at the bottom of the cooking olla before they added the pumpkin to be boiled.

Pumpkin blossoms can be prepared in the same ways as squash blossoms, although they were not utilized as much. (See "Squash".)

Here are some ways the various Indian groups prepared pumpkin:

NAVAJO PUMPKIN

Cut and soak dried pumpkin or cut fresh pumpkin into cubes. Boil until soft. In a frying pan fry tiny pieces of mutton and mutton fat. Add the cooked meat and the grease to the pumpkin and stir and mash all together. The proportion of meat to pumpkin varies from 1:1 to 1:8 depending on what foods are on hand.

APACHE PUMPKIN

1. Boil pumpkin pieces until soft and mash. Add salt to taste and 2 or 3 tablespoons ground sunflower seed or parched cornmeal for each cup of pumpkin.

2. Cut fresh pumpkin into thin slices and fry until soft in grease.[2]

HAVASUPAI PUMPKIN

Cut green corn from cob and mash or grind until pulpy. Cut pumpkin into cubes, about as much pumpkin as you have corn. Heat water to almost boiling and add corn, pumpkin, and salt. Cook and stir until vegetables are done. Cool before eating.

TAOS PUMPKIN
Yield: 4 servings

Cut kernels off 4 ears of sweet corn, chop fine 2 small unripe Indian pumpkins and half an onion. Fry all in a little fat in a frying pan, then cover the pan and steam until done.

150

NEW MEXICO PUMPKIN CANDY (Spanish influence)

Yield: 15–20 pieces

Put 1 quart of pumpkin cubes in a pan and cover with water. Bring to a boil and cook 15–20 minutes. Drain, reserving liquid. There should be about 1½ cups liquid. Mix the liquid with 1 cup brown sugar. Pour over pumpkin and boil 15 minutes. Let the pumpkin pieces stand in the syrup overnight. The next day boil the pumpkin in the syrup for 5 minutes. Take the pumpkin pieces from the syrup and set on waxed-paper lined tray in the sun to dry. After about 10 hours roll the pieces in granulated sugar. (The more coarse sugar that is purchased in Mexico is better than U.S. superfine.)

Pauline Goodmorning of Taos said to me, "I don't know how she did it, but my mother used to fry pumpkin seeds so they were just like nuts. We used to eat them at night when we were sitting around by the fire and the men were telling the stories." Unfortunately, we do not have the old recipe, but here is one method for preparing pumpkin seeds.

ROASTED PUMPKIN SEEDS

Remove seeds from pumpkin and dry in sun on a tray. Put seeds in a bowl and pour over them just enough vegetable oil to lightly coat them. Mix well. Add salt to taste and stir. Spread the oiled seeds on a metal tray and put in a 250° oven. Stir occasionally and roast until pumpkin seeds have begun to turn brown.

OTHER USES: The Cocopas rubbed the oily kernels of pumpkin seeds on their hands as protection against cold and a face paint was mixed with the oily seeds. This group also used pumpkin seeds to tan hides.[3]

A variety of pumpkin with a long, crooked neck was sometimes used as a container. It was necessary for the pumpkin to be completely mature before it was hollowed out or the wall would collapse. The neck portion was cut off for use as a cup and the glandular base was used as a jar.[4]

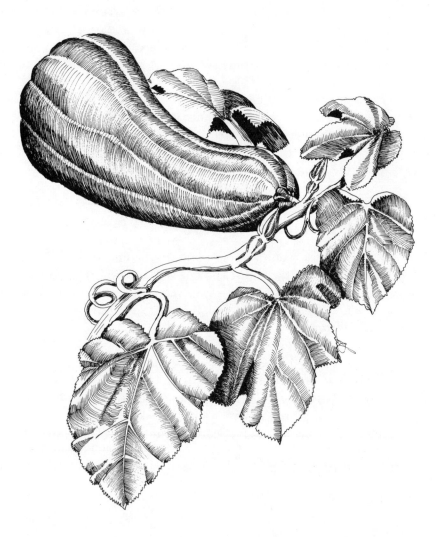

squash

Southwestern Indians raised a variety of squashes of both the summer and winter types. Evidence implies that some kinds were being grown before contact with the white man while other kinds were adopted after the arrival of the Spaniards.

Squashes were eaten from their first appearance when they were tiny and green all through their development until maturity. Squash was prepared for winter storage by removing the seeds and cutting it into strips, which were sun-dried. The seeds were also sun-dried and used.

Besides baking and boiling squash as we do today, Indians prepared this vegetable and its male blossoms in the following ways.

HAVASUPAI SQUASH-BLOSSOM PUDDING

Yield: 4–5 servings

3 ears of green corn
2 or 3 cups squash blossoms
salt

Cut green corn from cob and cook in water over medium heat for about 30 minutes. Wash squash blossoms and remove stem. Boil blossoms until tender, then mash to a pulp. Add mashed blossoms to green corn and cook until thick. Season with salt.

FRIED SQUASH BLOSSOMS

½ cup flour
1 cup milk
dash of salt
½ teaspoon chili powder
oil
1 quart squash blossoms

Mix together flour, milk, salt, and chili powder. Heat oil in frying pan. Dip blossoms in batter and fry in hot oil until crisp.

YUMAN SQUASH

Clean and pare fresh squash. Cut into pieces and boil until soft. Drain and mash. Heat animal fat in pan or pot and add squash. Stir until thoroughly mixed. Season with salt.

Present-day Pueblo Indians prepare this squash dish:

CALABACITAS

Yield: 4–6 servings

3 ears corn
1 onion
fresh garlic
summer squash
2 tomatoes
crumbled cheese

Cut kernels off ears of corn and slice onion and mince garlic. Combine with a little water in a covered pan. Cook 5 minutes. Add sliced squash and cook another 10 minutes. Add sliced tomatoes and cook until tomatoes are warm but not mushy. Add crumbled yellow or white cheese. Mix lightly and serve.

Mexicans make a squash-seed gravy called Pipian. Here is one recipe for this tasty sauce:

PIPIAN
Yield: 3–4 cups sauce

3 tablespoons lard
2 slices dry bread
2 dried red chilis
1 clove garlic
1 cup dried squash seeds
1½ tablespoons flour
sugar to taste
3 cups water

Heat lard in frying pan. Break up the dry bread to small pieces, and tear up the chilis, discarding the seeds. Chop garlic. Add bread, chilis, garlic, and squash seeds to the hot lard and cook until bread and seeds are toasted. Remove solids from the grease and grind fine. Return powder to the lard along with the flour, sugar, and water. Cook until thick, stirring constantly. Use as a sauce for chicken or vegetables.

OTHER USES: Squashes of the long variety were worn as phallic symbols in dances. Squashes with hard rinds were sometimes hollowed out and used as receptacles for storage or as water dippers.[1]

wheat

Although wheat was late arriving among the Indians, many groups took to this new grain readily and among some, such as the Zunis and Pimas, it soon outranked corn in importance. In the revolt against the Spaniards the Tewa groups burned all their wheat, despising it as an importation, but now wheat is highly valued and ritually mentioned along with the aboriginal foodstuffs. It has even been introduced into stories that purport to describe pre-Spanish events.[1]

Indian women devised several methods of making yeast so that their wheat loaves would rise and become light. Very early Pueblo women used the green parasitic lichen (lungwort), which is found growing on hardwood trees, steeping it in warm water overnight and mixing it with flour in the morning.[2]

When using cornmeal as part of the ingredient for wheat bread, a Pueblo woman would sometimes mix the cornmeal with water and put some

155

An Indian woman at Isleta Pueblo in central New Mexico grinds corn the way her ancestors have done for more than a thousand years. The different troughs are used to grind the corn or wheat to successively finer stages. (*Photo courtesy New Mexico Department of Development.*)

in her mouth and chew it. When the chewed lumps were returned to the dough the enzymes of the saliva with the bacteria caused fermentation of the whole mass.

The usual method among the Zunis was to make a sour dough by mixing up a sponge with flour and water and letting it stay over night. When a successful sponge had been started, a bit of the dough was saved from one baking to the next. The sponge dough usually hardened and the baker either ground it into powder and mixed it with water or soaked the lump in water.

Cleaning the ashes from her adobe beehive oven or *horno*, a Pueblo Indian woman readies it for baking fragrant, tasty loaves of wheat bread. (*Arizona State Museum—University of Arizona, Forman Hanna Collection.*)

156

The most primitive method of baking this dough was to bury it in ashes. It was also made into flat cakes and toasted for tortillas or fried in hot lard for the ubiquitous Indian fry bread. But the most popular method of baking bread among the Zuni and Pueblo groups was in their adobe beehive ovens. (See sources section at end of book for places to obtain plans for building an adobe oven.)

Yuman tribes typically prepared wheat mush and unleavened tortillas from their grain, which was always parched before being ground. Sometimes parched flour was simply eaten in pinches with water. One of the few beverages used among the groups living along the lower Colorado River was made from thoroughly ripened wheat grains that were lightly roasted over charcoal, then ground to a fine flour and mixed with water. This pinole was greatly relished. Sometimes the mixture was allowed to ferment and produced a mildly intoxicating drink.[3]

PIMA ASH BREAD
Yield: 1 small loaf

Form a stiff dough of hand-ground wheat, salt, and water. Shape dough into a cake 3 inches thick in the middle and thinner at the edges.

Build a fire of mesquite wood. When it is reduced to coals, push them aside and make a depression in the hot ashes with a bowl. Place the loaf in the hollow, cover with ashes and hot coals and bake about 1 hour. Test the bread, and if it is done, brush off the ashes. Pimas used a corn cob for this purpose.[4]

Fry bread is made by, the Apaches, Pimas, Papagos, Zunis, Hopis, Navajos, and the peoples of the Rio Grande pueblos, among other groups:

FRY BREAD
Yield: 12–14 bread puffs

4 cups flour (all white or ½ white and ½ whole wheat)
1 tablespoon baking powder
1½ teaspoons salt
1½ teaspoons cooking oil
water

Mix the flour, baking powder, and salt. Add oil and enough water to made a soft dough. Knead with a floured hand until the dough has some spring. Shape into balls about 3 inches in diameter. Pat each ball into a flat cake and keep patting and stretching the dough into a thin sheet, 10 to 12 inches in diameter. (This takes practice.) Fry each round of dough in very hot fat, turning once, until puffy and golden.

When Loretta Blatchford, a very good Navajo cook, demonstrated the above recipe for me, she confessed that she usually substitutes 1 tablespoon of mayonnaise for the oil in the traditional recipe.

INDIAN FLOUR TORTILLAS

Yield: 10 tortillas

2 cups flour
1 teaspoon salt
¼ cup lard
½ cup lukewarm water

Combine the flour and salt in a bowl. Using hands, squeeze the fat into the flour until all the particles of lard are very small. Gradually add the water, mixing and kneading until the dough is smooth and elastic. Divide the dough into 10 balls. Pat the balls flat and pat and stretch (resort to a rolling pin if necessary) until the dough is as thin as you can get it.

Drop each tortilla onto a very hot ungreased griddle or frying pan. Bake until freckled on one side and turn to cook the other side.

This is a staple in many Indian households and is eaten at all meals.

The following recipe was borrowed from the Spaniards but now is made in many Pueblo Indian homes.

PANOCHA

Yield: 6–8 servings

2 cups sprouted wheat meal
1 piloncillo (pyramid-shaped brown sugar, or substitute ⅓ cup brown sugar)
2 cups hot water
1 cup white wheat flour

Purchase sprouted wheat meal (panocha flour) at grocery store or prepare as follows: Put wheat in a cloth sack or jar to soak for 24 hours, drain, and spread grain out on clean dish towels to sprout. Dry the sprouted grain, grind as flour, and sift.

Put piloncillo in bowl and cover with 1 cup boiling water. Mix flour and sprouted wheat meal. When the piloncillo has dissolved in the water, add this syrup and 1 more cup of hot water to the flour and stir to make a very thick batter. Let the batter stand 1 hour then pour into a flat, greased 1 quart casserole or oven-proof glass pan. Bake for 2 hours in a 250° oven until the pudding separates from the side of the pan.

This can be served as a side dish with a meal or as a dessert with the addition of a sweet topping. If it is used as a dessert, I prefer to double the amount of sugar in the recipe.

The first time I ate the following dish was during an amazing afternoon. I was in northern New Mexico doing early research for this book and one day decided to walk into Taos to see what the town library had to offer on the subject of Indian foods. As I was strolling along the dirt road an old pickup stopped beside me. It was driven by an older Indian man with a beautiful face and long braids done up with ribbons. His passenger was a young woman about my age. The driver offered me a ride and introduced himself as Tellus Goodmorning. I had heard of this man, as he was a great favorite and friend of the many young people who had moved to the Taos area. As we drove toward town he told me that he was taking the other woman to the Fiesta of San Lorenzo at Picuris pueblo and asked if I would like to go along. What an alternative to spending the afternoon in the library!

We drove over the lush green mountains separating Taos and Picuris and Tellus reminisced on how he had often made the trip on horseback as a child. When we arrived, the dances were recessed for lunch. That was my introduction to the Pueblo custom of each household preparing mountains of food during fiestas to serve to any friends coming from other villages for the dances.

Tellus took me and his other guest to the home of a distant relative, Reycita Vanijo. As we entered the small front room, I could see people seated around a large table in the kitchen. They were eating, and several women, very old to quite young, waited on them, bringing bowls of food to the table. When these people had finished eating, the table was cleared, the dishes were washed, and the table reset. The three of us, along with some other persons who had gathered in the room, were then invited to eat.

I was surprised, humbled, and enormously grateful to be included in the invitation. And what a feast they had laid! We had meat and potatoes in chili gravy, cabbage cooked with chunks of beef, fruit salad, garbanzo beans (chick peas) cooked in beef broth, meatloaf, bread baked in the adobe oven, gelatin dessert, the sweet-bread pudding below, and very thin squares of apple pie, with cold fruit punch and coffee to drink. At a feast I attended at the Santa Clara pueblo a year later, also with Tellus, the household where we ate offered much the same variety of food—heaping plates of food, all delicious. The day after the outing to the Picuris fiesta Tellus introduced me to his wife Pauline, who gave me the following recipe:

PUEBLO SWEET BREAD PUDDING (Sopa)
Yield: 8–10 servings

1 loaf white bread (homemade adobe bread is best)
½ pound longhorn cheese
1 cup raisins
2 cups sugar
hot water
1 teaspoon vanilla
2 shakes of cloves
2 shakes of cinnamon

Slice the loaf of bread and toast very dry but do not burn. Break the toast into little pieces and put into a pan, layered with thin slices of cheese and raisins.

Brown the sugar in a medium-sized frying pan, stirring constantly until it melts and caramelizes. Slowly add enough hot water to almost fill the frying pan. Add the vanilla and spices. Let the mixture boil for 10 minutes and then pour the syrup over the bread in the pan. Bake the pudding in a 300° oven until the cheese is melted and all the liquid is absorbed.

Pauline maintains that this dish is best served a day old and warmed up.

ADOBE OVEN BREAD
Yield: 2 9-inch round loaves

1 package yeast
½ cup warm water
5 cups (or more) all-purpose flour
2 tablespoons shortening
1½ cups hot water
2 teaspoons salt
2 teaspoons sugar

Dissolve yeast in warm water. Combine shortening, hot water, salt, and sugar in large bowl. Add 1 cup of flour and beat. Add yeast and beat again. Add three more cups of flour. Spread the remaining cup of flour on a board or working surface and turn the dough on to it. Spread some of the flour over the sticky dough and knead for 10 minutes, adding more flour as needed to keep the dough from sticking to the board. The total amount of flour needed will depend on how much moisture is in the flour when you begin.

Grease a bowl and put the elastic ball of dough in, turning once to coat the top with grease. Let rise in a warm place until doubled in bulk. Divide dough into 2 equal portions. Shape into round loaves.

For special occasions the Pueblo women often fashion animals from the dough. You can make a loaf that looks something like a bird by rolling each ball of dough to make a circle 8 inches in diameter. Fold the circle almost in half with the bottom half of the circle protruding about 1 inch from under the top half. Now divide the dough into 3 sections by cutting ⅔ of the way into the loaf, starting at the curved edge. Put each loaf in a greased 9-inch pie pan, spreading the sections apart a bit and letting the outer edges extend up the rim of the pan.

Cover the dough and let rise again until doubled. Bake at 350° for about 1 hour. A pan of water placed in the oven, but not directly under the bread, will help to make a crunchy crust.

Additions: The Hopis and Zunis sometimes mixed little pieces of meat or vegetables with the dough or included a meat and vegetable mixture in a pocket in the middle of the bread.

At times, ground seeds or nuts of various types were used as part of the flour.

OTHER USES: Tewa tribes used wheat gruel as a remedy for stomach disorders and diarrhea.[5]

The Pimas believed that green wheat in the "milk" stage was good for encouraging mother's milk. A thin porridge was made from wheat just beyond the "milk" stage, ground saguaro seeds and cholla buds all mixed with water and boiled.[6]

Notes

INTRODUCTION

1. By permission from *A Pima Remembers*, by George Webb, Tucson, University of Arizona Press, copyright 1959.

2. Kent V. Flannery, "Archeological Systems Theory and Early Mesoamerica," *Anthropological Archeology in the Americas* (Washington, D.C., 1968), p. 67.

3. *Ibid.*, p. 69.

4. Katharine Bartlett, "Edible Wild Plants of Northern Arizona," *Plateau*, Vol. 16, No. 1 (July 1943), p. 11.

5. Matilda Cox Stevenson, "Ethnobotany of the Zuni," *30th Annual Report of the Bureau of American Ethnology* (Washington, D.C., 1909), p. 36.

6. Keith H. Basso, *The Cibecue Apache* (New York, 1970), p. 3.

7. Display card at Museum of Northern Arizona, Flagstaff, Arizona.

8. H. M. Wormington, *Prehistoric Indians of the Southwest*, 3rd Ed. (Denver, 1956), pp. 40–45.

9. James B. Watson, "How the Hopi Classify Their Foods," *Plateau*, Vol. 15, No. 4 (April 1943), pp. 49–51.

10. Leslie Spier, *Yuman Tribes of the Gila River* (Chicago, 1933), pp. 78–81.

11. Frank Russell, "The Pima Indians," *26th Annual Report of the Bureau of American Ethnology* (Washington, D.C., 1908), pp. 66–67.

12. Edward Castetter and Ruth Underhill, "Ethnobiological Studies in the American Southwest II: Ethnobiology of the Papago Indians," *University of New Mexico Bulletin*, Vol. 4, No. 3, Biological Series, 1935, pp. 45–46.

13. Flora L. Bailey, "Navajo Foods and Cooking Methods." *American Anthropologist*, Vol. 42, No. 2 (April–June 1940), pp. 273–79.

14. Wilfred W. Robbins, John Peabody Harrington, and Barbara Freire-Marreco, *Ethnobotany of the Tewa Indians*, Smithsonian Institution Bureau of American Ethnology Bulletin 55 (Washington, D.C., 1916) pp. 94–95.

15. *Ibid.*

16. Edmund Nequatewa, "Hopi Courtship and Marriage," *Museum Notes*, Vol. 5, No. 9 (March 1933), p. 49.

17. Robbins, Harrington, and Freire-Marreco, *Ethnobotany of the Tewa Indians*, p. 77.

I. CACTUS AND CACTUSLIKE PLANTS

Agave

1. Edward Castetter, Willis H. Bell, and M. E. Opler, "Ethnobiological Studies in the American Southwest III: The Ethnobiology of the Chiricahua and Mescalero Apache," *University of New Mexico Bulletin*, Vol. 4, No. 5, Biological Series, 1936, p. 35.

2. *Ibid.*, pp. 36–37.

3. Alfred Whiting, *Ethnobotany of the Hopi* (Flagstaff, 1966), p. 71.

4. Julia F. Morton, "Principal Wild Plant Foods of the United States excluding Alaska and Hawaii," *Economic Botany*, Vol. 17, No. 4 (October–December 1963), p. 370.

5. Winifred Ross, *The Present Day Dietary Habits of the Papago Indians*, unpublished master's thesis for the University of Arizona, 1941, pp. 43–44.

6. Edward Castetter, Willis Bell, and Alvin R. Grove, "Ethnobiological Studies in the American Southwest VI: The Early Utilization and the Distribution of Agave in the American Southwest,"*University of New Mexico Bulletin*, Vol. 5, No. 4, Biological Series, 1938, pp. 64–76.

7. Winfred Buskirk, *Western Apache Subsistence Economy*, unpublished doctoral dissertation for the University of New Mexico, 1949, p. 306.

8. Castetter, Bell, and Grove, "Utilization and Distribution of Agave," pp. 64–76.

9. Whiting, *Ethnobotany of the Hopi*, p. 71.

10. Castetter, Bell, and Grove, "Utilization and Distribution of Agave," pp. 76–77.

11. Ignaz Pfefferkorn, *Sonora, a Description of the Province* (Albuquerque, 1949), p. 60.

Barrel Cactus

1. Natt N. Dodge, Flowers of the Southwest Deserts (Globe, Arizona, 1961), p. 60.

2. L. S. M. Curtin, *By the Prophet of the Earth*, (Santa Fe, 1949), p. 56.

3. *Ibid.*

Cholla

1. Alfred Whiting, *Ethnobotany of the Hopi* (Flagstaff, 1966), p. 86.
2. Frank Russell, "The Pima Indians," *26th Annual Report of the Bureau of American Ethnology* (Washington, D.C., 1908), p. 71.
3. Edmund Nequatewa, "Some Hopi Recipes For the Preparation of Wild Plant Foods," *Plateau*, Vol. 16, No. 1 (July 1943), pp. 18–19.
4. Winifred Ross, *The Present Day Dietary Habits of the Papago Indians*, unpublished master's thesis for the University of Arizona, 1941, p. 43.
5. L. S. M. Curtin, *By the Prophet of the Earth* (Santa Fe, 1949), pp. 58–59.
6. Whiting, *Ethnobotany of the Hopi*, p. 86.
7. Winfred Buskirk, *Western Apache Subsistence Economy*, unpublished doctoral dissertation for the University of New Mexico, 1949, p. 322.
8. A. Knochmal, S. Paur, and P. Duisberg, "Useful Native Plants in the American Southwest Deserts," *Economic Botany*, Vol. 8, No. 1 (January–March 1954), p. 6.

Ocotillo

1. L. S. M. Curtin, *By the Prophet of the Earth* (Santa Fe, 1949), p. 90.
2. Ignaz Pfefferkorn, *Sonora, a Description of the Province* (Albuquerque, 1949), p. 66.

Prickly Pear

1. Flora L. Bailey, "Navajo Foods and Cooking Methods," *American Anthropologist*, Vol. 42, No. 2 (April–June, 1949), p. 278.
2. Frank Russell, "The Pima Indians," *26th Annual Report of the Bureau of American Ethnology* (Washington, D.C., 1908), p. 76.
3. Leslie Spier, *Yuman Tribes of the Gila River* (Chicago, 1933), p. 54.
4. Edward Castetter, Willis H. Bell, and M. E. Opler, "Ethnobiological Studies in the American Southwest III: The Ethnobiology of the Chiricahua and Mescalero Apache," *University of New Mexico Bulletin*, Vol. 4, No. 5, Biological Series, 1936, p. 40.
5. Winifred Ross, *The Present Day Dietary Habits of the Papago Indians*, unpublished master's thesis for the University of Arizona, 1941, pp. 43–44.
6. L. S. M. Curtin, *By the Prophet of the Earth* (Santa Fe, 1949), p. 61.
7. L. S. M. Curtin, *Healing Herbs of the Upper Rio Grande* (Los Angeles, 1965), p. 134.
8. Donald Kirk, *Wild Edible Plants of the Western United States* (Healdsburg, California, 1970), p. 50.

Saguaro

1. Jacob Baeget, "The Aboriginal Inhabitants of the California Peninsula," *Annual Report of the Board of Regents of the Smithsonian Institution* (Washington, D.C., 1863), p. 363.
2. Winifred Ross, *The Present Day Dietary Habits of the Papago Indians*, unpublished master's thesis for the University of Arizona, 1941, pp. 43–44.
3. Edward Castetter and Willis H. Bell, "Ethnobiological Studies in the American Southwest IV: The Aboriginal Utilization of the Tall Cacti in the American Southwest," *University of New Mexico Bulletin*, Vol. 5, No. 1, Biological Series, 1937, pp. 22–23.
4. Leslie Spier, *Yuman Tribes of the Gila River* (Chicago, 1933) pp. 57–58.
5. Frank Russell, "The Pima Indians," *26th Annual Report of the Bureau of American Ethnology* (Washington, D. C., 1908), p. 26.

164

6. L. S. M. Curtin, *By the Prophet of the Earth* (Santa Fe, 1949), p. 53.
7. *Ibid.*

Yucca: Datil and Palmilla

1. Willis H. Bell and Edward F. Castetter, "Ethnobiological Studies in the American Southwest VII: The Utilization of Yucca, Sotol and Beargrass by the Aborigines in the American Southwest," *University of New Mexico Bulletin*, Vol. 5, No. 5, Biological Series, 1941, p. 65.

2. Wilfred W. Robbins, John Peabody Harrington, and Barbara Freire-Marreco, *Ethnobotany of the Tewa Indians*, Smithsonian Institution Bureau of American Ethnology Bulletin 55 (Washington D.C., 1916), p. 65.

3. Bell and Castetter, "Utilization of Yucca," p. 56.

4. *Ibid.*, p. 55.

5. *Ibid.*, p. 56.

6. Robbins, Harrington, and Freire-Marreco, *Ethnobotany of the Tewa Indians*, p. 51.

7. Edward Castetter, Willis H. Bell, and M. E. Opler, "Ethnobiological Studies in the American Southwest III: The Ethnobiology of the Chiricahua and Mescalero Apache," *University of New Mexico Bulletin*, Vol. 4, No. 5, Biological Series, 1936, p. 40.

8. Matilda Cox Stevenson, "Ethnobotany of the Zuni," *30th Annual Report of the Bureau of American Ethnology* (Washington, D.C., 1909), p. 75.

9. Castetter, Bell, and Opler, "Ethnobiology of the Chiricahua and Mescalero Apache," p. 39.

10. Donald Kirk, *Wild Edible Plants of the Western United States* (Healdsburg, California, 1970), p. 279.

11. Bell and Castetter, "Utilization of Yucca," pp. 22–54 *passim*.

12. Ibid., p. 52.

13. Robbins, Harrington, and Freire-Marreco, *Ethnobotany of the Tewa Indians*, p. 52.

14. Alfred Whiting, *Ethnobotany of the Hopi* (Flagstaff, 1966), p. 71.

15. A. Kochmal, S. Paur, and P. Duisberg, "Useful Native Plants in the American Southwest Deserts," *Economic Botany*, Vol. 8, No. 1 (January–March, 1954), p. 4.

II. NUTS AND SEEDS

Acorn

1. J. I. Lighthall, *The Indian Household Medecine Guide*, 2nd Ed., (Mokelumne Hill, California, 1966), p. 50.

Grass Seed

1. Edward Castetter and Ruth Underhill, "Ethnobiological Studies in the American Southwest: The Ethnobiology of the Papago Indians," *University of New Mexico Bulletin*, Vol. 4, No. 3, Biological Series, 1935, p. 24.

2. Edward Castetter and Willis H. Bell, *Yuman Indian Agriculture* (Albuquerque, 1951), p. 188.

3. Edward Castetter, Willis H. Bell, and M. E. Opler, "Ethnobiological Studies in the American Southwest III: The Ethnobiology of the Chiricahua and Mescalero Apache," *University of New Mexico Bulletin*, Vol. 4, No. 5, Biological Series, 1936, p. 49.

Jojoba

1. Unknown Jesuit padre, *Rudo Ensayo* (Tucson, 1951), p. 46.
2. Ignaz Pfefferkorn, *Sonora, a Description of the Province* (Albuquerque, 1949), p. 65.

Mesquite

1. Edward Castetter and Willis H. Bell, *Yuman Indian Agriculture* (Albuquerque, 1951), pp. 179–180.
2. Richard S. Felger and Mary B. Moser, "Seri Use of Mesquite," *The Kiva*, Vol. 37, No. 1 (Fall 1971), p. 55.
3. Winifred Ross, *The Present Day Dietary Habits of the Papago Indians*, unpublished master's thesis for the University of Arizona, 1941, pp. 43–44.
4. Castetter and Bell, *Yuman Indian Agriculture*, p. 183.
5. Leslie Spier, *Yuman Tribes of the Gila River* (Chicago, 1933), p. 52.
6. Thomas B. Hinton, "A Description of the Contemporary Use of an Aboriginal Sonoran Food," *The Kiva*, Vol. 21, Nos. 3 and 4, pp. 27–28.
7. L. S. M. Curtin, *By the Prophet of the Earth* (Santa Fe, 1949), p. 95.
8. Frank Russell, "The Pima Indians," *26th Annual Report of the Bureau of American Ethnology* (Washington, D.C., 1908), p. 79.
9. L. S. M. Curtin, *Healing Herbs of the Upper Rio Grande* (Los Angeles, 1965), p. 191.
10. Russell, "The Pima Indians," p. 79.
11. Castetter and Bell, *Yuman Indian Agriculture*, p. 183.
12. Curtin, *By the Prophet of the Earth*, p. 94.

Pinyon Pine

1. Wilfred W. Robbins, John Peabody Harrington, and Barbara Freire-Marreco, *Ethnobotany of the Tewa Indians*, Smithsonian Institution Bureau of American Ethnology Bulletin 55 (Washington, D. C., 1916), p. 41.
2. Edward Castetter, Willis H. Bell, and M. E. Opler, "Ethnobiological Studies in the American Southwest III: The Ethnobiology of the Chiricahua and Mescalero Apache," *University of New Mexico Bulletin*, Vol. 4, No. 5, Biological Series, 1936, p. 43.
3. Winfred Buskirk, *Western Apache Subsistence Economy*, unpublished doctoral dissertation for the University of New Mexico, 1949, p. 332.
4. Alfred Whiting, *Ethnobotany of the Hopi* (Flagstaff, 1966), p. 63.
5. Matilda Cox Stevenson, "Ethnobotany of the Zuni," *30th Annual Report of the Bureau of American Ethnology* (Washington, D.C., 1909), p. 58.

Sunflower

1. Alfred Whiting, *Ethnobotany of the Hopi* (Flagstaff, 1966), p. 97.
2. L. S. M. Curtin, *By the Prophet of the Earth* (Santa Fe, 1949), p. 104.
3. Wilfred W. Robbins, John Peabody Harrington, and Barbara Freire-Marreco, *Ethnobotany of the Tewa Indians*, Smithsonian Institution Bureau of American Ethnology Bulletin 55 (Washington, D.C., 1916), p. 56.
4. Matilda Cox Stevenson, "Ethnobotany of the Zuni," *30th Annual Report of the Bureau of American Ethnology* (Washington, D.C., 1909), p. 58.
5. Edward Castetter and Willis H. Bell, *Yuman Indian Agriculture*, (Albuquerque, 1951), p. 196.
6. Ben Charles Harris, *Eat the Weeds* (Barre, Massachusetts, 1971), p. 203.
7. Pauline M. Patraw, *Flowers of the Southwest Mesas* (Globe, Arizona, 1964), p. 81.
8. Curtin, *By the Prophet of the Earth*, p. 104.

9. *Ibid.*

10. Harris, *Eat the Weeds*, p. 199.

11. Muriel Sweet, *Common Edible and Useful Plants of the West* (Healdsburg, California, 1962), p. 59.

Black Walnut

1. Irma S. Rombauer and Marion Rombauer Becker, *The Joy of Cooking* (Indianapolis, 1964), p. 519.

2. Winfred Buskirk, *Western Apache Subsistence Economy*, unpublished doctoral dissertation for the University of New Mexico, 1949, p. 136.

3. L. S. M. Curtin, *Healing Herbs of the Upper Rio Grande* (Los Angeles, 1965), p. 132.

III. GRAPES, BERRIES, AND CHERRIES

Chokecherry

1. Mrs. J. H. Watt, "Chokecherry Wine," *Cooking in Wyoming, 1890–1965* (Cheyenne, 1965), p. 78.

2. L. S. M. Curtin, *Healing Herbs of the Upper Rio Grande* (Los Angeles, 1965), p. 52.

Wild Currant

1. Edmund Nequatewa, "Some Hopi Recipes for the Preparation of Wild Plant Foods," *Plateau*, Vol. 16, No. 1 (July 1943), p. 18.

2. Matilda Cox Stevenson, "Ethnobotany of the Zuni," *30th Annual Report of the Bureau of American Ethnology* (Washington, D.C., 1909), p. 70.

3. Thomas H. Kearney and Robert H. Peebles, *Arizona Flora* (Berkeley, and Los Angeles, 1969), p. 368.

Elderberry

1. L. S. M. Curtin, *Healing Herbs of the Upper Rio Grande* (Los Angeles, 1965), p. 88.

2. L. S. M. Curtin, *By the Prophet of the Earth* (Santa Fe, 1949), p. 75.

3. *Ibid.*

4. J. I. Lighthall, *The Indian Household Medicine Guide*, 2nd Ed., (Mokelumne Hill, California, 1966), p. 65.

5. Euell Gibbons, *Stalking the Wild Asparagus* (New York, 1962), p. 94.

Wild Grape

1. Thomas H. Kearney and Robert H. Peebles, *Arizona Flora* (Berkeley and Los Angeles, 1969), p. 535.

2. Donald Kirk, *Wild Edible Plants of the Western United States* (Healdsburg, California, 1970), p. 264.

Ground Cherry

1. L. S. M. Curtin, *By the Prophet of the Earth* (Santa Fe, 1949), p. 88.

2. Edward Castetter and Willis H. Bell, *Yuman Indian Agriculture,* (Albuquerque, 1951), p. 207.

3. Wilfred W. Robbins, John Peabody Harrington, and Barbara Freire-Marreco, *Ethnobotany of the Tewa Indians*, Smithsonian Institution Bureau of American Ethnology Bulletin 55 (Washington, D.C., 1916), p. 59.

Manzanita

    1. L. S. M. Curtin, *Healing Herbs of the Upper Rio Grande* (Los Angeles, 1965), p. 68.

Rose

    1. Wilfred W. Robbins, John Peabody Harrington, and Barbara Freire-Marreco, *Ethnobotany of the Tewa Indians*, Smithsonian Institution Bureau of American Ethnology Bulletin 55 (Washington, D.C., 1916), p. 48.

    2. L. S. M. Curtin, *Healing Herbs of the Upper Rio Grande* (Los Angeles, 1965), pp. 172–73.

    3. Muriel Sweet, *Common Edible and Useful Plants of the West* (Healdsburg, California, 1962), p. 20.

Squawberry

    1. Edward Castetter, Willis H. Bell, and M. E. Opler, "Ethnobiological Studies in the American Southwest III: The Ethnobiology of the Chiricahua and Mescalero Apache," *University of New Mexico Bulletin*, Vol. 4, No. 5, Biological Series, 1936, p. 35.

    2. Elisabeth Hart, *Pima Cookery*, 2nd Ed. (Sacaton, Arizona, 1968), p. 12.

    3. Alfred Whiting, *Ethnobotany of the Hopi* (Flagstaff, 1966), p. 84.

    4. L. S. M. Curtin, *Healing Herbs of the Upper Rio Grande* (Los Angeles, 1965), p. 112.

    5. Whiting, *Ethnobotany of the Hopi*, p. 84.

    6. *Ibid.*

    7. *Ibid.*

Wolf Berry

    1. Matilda Cox Stevenson, "Ethnobotany of the Zuni," *30th Annual Report of the Bureau of American Ethnology* (Washington, D.C., 1909), p. 94.

    2. *Ibid.*, p. 68.

    3. Alfred Whiting, *Ethnobotany of the Hopi* (Flagstaff, 1966), p. 89.

    4. L. S. M. Curtin, *By the Prophet of the Earth* (Santa Fe, 1949), p. 88.

    5. Whiting, *Ethnobotany of the Hopi*, p. 89.

IV. FOODS OF MARSH AND MESA

Buffalo Gourd

    1. A. Knockmal, S. Paur, and P. Duisberg, "Useful Native Plants in the American Southwest Deserts," *Economic Botany*, Vol. 8, No. 1 (January–March 1954), p. 8.

    2. L. S. M. Curtin, *Healing Herbs of the Upper Rio Grande* (Los Angeles, 1965), p. 47.

    3. Frank Russell, "The Pima Indians," *26th Annual Report of the Bureau of American Ethnology* (Washington, D.C., 1908), p. 79.

    4. Wilfred W. Robbins, John Peabody Harrington, and Barbara Freire-Marreco, *Ethnobotany of the Tewa Indians*, Smithsonian Institution Bureau of American Ethnology Bulletin 55 (Washington, D.C., 1916), p. 63.

    5. Winfred Buskirk, *Western Apache Subsistence Economy*, unpublished doctoral dissertation for the University of New Mexico, 1949, p. 345.

    6. Muriel Sweet, *Common Edible and Useful Plants of the West* (Healdsburg, California, 1962), p. 7.

1. Ben Charles Harris, *Eat the Weeds* (Barre, Massachusetts, 1971), pp. 88–89.

2. L. S. M. Curtin, *By the Prophet of the Earth* (Santa Fe, 1949), p. 64.

3. James E. Churchill, "Food Without Farming," *Mother Earth News*, Vol. 1, No. 3 (May 1970), pp. 59–60.

4. Alfred Whiting, *Ethnobotany of the Hopi* (Flagstaff, 1966), p. 39.

5. *Ibid.*, p. 65.

6. *Ibid.*, p. 64.

Cota

1. Thomas H. Kearney and Robert H. Peebles, *Arizona Flora*, 2nd Ed., (Berkeley and Los Angeles, 1969), p. 909.

Devil's Claw

1. Alfred Whiting, *Ethnobotany of the Hopi* (Flagstaff, 1966), p. 92.

2. A. Knochmal, S. Paur, and P. Duisberg, "Useful Native Plants in the American Southwest Deserts," *Economic Botany*, Vol. 8, No. 1 (January–March 1954), p. 8.

3. L. S. M. Curtin, *By the Prophet of the Earth* (Santa Fe, 1949), p. 108.

Mormon Tea

1. Donald Kirk, *Wild Edible Plants of the Western United States*, (Healdsburg, California, 1970), p. 21.

2. L. S. M. Curtin, *Healing Herbs of the Upper Rio Grande* (Los Angeles, 1965), p. 59.

3. L. S. M. Curtin, *By the Prophet of the Earth* (Santa Fe, 1949), p. 76.

4. Muriel Sweet, *Common Edible and Useful Plants of the Southwest* (Healdsburg California, 1962), p. 16.

5. Curtin, *By the Prophet of the Earth*, p. 76.

6. Thomas H. Kearney and Robert H. Peebles, *Arizona Flora* (Berkeley and Los Angeles, 1969), p. 60.

7. Sweet, *Common Edible and Useful Plants*, p. 16.

Puffball

1. Matilda Cox Stevenson, "Ethnobotany of the Zuni," *30th Annual Report of the Bureau of American Ethnology* (Washington, D.C., 1909), p. 69.

2. Wilfred W. Robbins, John Peabody Harrington, and Barbara Freire-Marreco, *Ethnobotany of the Tewa Indians*, Smithsonian Institution Bureau of American Ethnology Bulletin 55 (Washington, D.C., 1916), p. 66.

3. L. S. M. Curtin, *Healing Herbs of the Upper Rio Grande* (Los Angeles, 1965), p. 100.

V. GREENS

Beeweed

1. L. S. M. Curtin, *Healing Herbs of the Upper Rio Grande* (Los Angeles, 1965), p. 94.

2. Wilfred W. Robbins, John Peabody Harrington, and Barbara Freire-Marreco, *Ethnobotany of the Tewa Indians*, Smithsonian Institution Bureau of American Ethnology Bulletin 55 (Washington, D.C., 1916), p. 59.

3. L. S. M. Curtin, *Healing Herbs*, pp. 93–94.

4. Robbins, Harrington, and Freire-Marreco, *Ethnobotany of the Tewa Indians*, p. 59.

Canaigre

1. L. S. M. Curtin, *By the Prophet of the Earth* (Santa Fe, 1949), p. 52.

2. Elisabeth Hart, *Pima Cookery*, 2nd Ed. (Sacaton, Arizona, 1968), p. 13.

3. Curtin, *By the Prophet of the Earth*, p. 52.

4. *Ibid.*

5. Matilda Cox Stevenson, "Ethnobotany of the Zuni," *30th Annual Report of the Bureau of American Ethnology* (Washington, D.C., 1909), p. 59.

6. Curtin, *By the Prophet of the Earth*, p. 52.

Curly Dock

1. L. S. M. Curtin, *By the Prophet of the Earth* (Santa Fe, 1949), p. 51.

2. *Ibid.*

3. Larry D. Olsen, *Outdoor Survival Skills* (Provo, Utah, 1973), p. 80.

4. L. S. M. Curtin, *Healing Herbs of the Upper Rio Grande* (Los Angeles, 1965), p. 114.

Dandelion

1. H. E. Bravery, *Successful Wine Making At Home* (New York, 1961), pp. 93–94.

2. Wilfred W. Robbins, John Peabody Harrington, and Barbara Freire-Marreco, *Ethnobotany of the Tewa Indians*, Smithsonian Institution Bureau of American Ethnology Bulletin 55 (Washington, D.C., 1916), p. 61.

3. *Ibid.*

4. L. S. M. Curtin, *Healing Herbs of the Upper Rio Grande* (Los Angeles, 1965), p. 60.

5. *Ibid.*, p.61.

6. J. I. Lighthall, *The Indian Household Medicine Guide*, 2nd Ed. (Mokelumne Hill, California, 1966), p. 53.

Lamb's-quarter

1. Alfred Whiting, *Ethnobotany of the Hopi* (Flagstaff, 1966), p. 74.

2. Muriel Sweet, *Common Edible and Useful Plants of the West* (Healdsburg, California, 1962), p. 38.

Horsemint

1. Wilfred W. Robbins, John Peabody Harrington, and Barbara Freire-Marreco, *Ethnobotany of the Tewa Indians*, Smithsonian Institution Bureau of American Ethnology Bulletin 55 (Washington, D.C., 1916), p. 57.

Pigweed

1. Matilda Cox Stevenson, "Ethnobotany of the Zuni," *30th Annual Report of the Bureau of American Ethnology* (Washington, D.C., 1909), p. 65.

2. Edward Castetter and Willis H. Bell, *Yuman Indian Agriculture* (Albuquerque, 1951), p. 189.

Monkey Flower

1. Muriel Sweet, *Common Edible and Useful Plants of the West* (Healdsburg, California, 1962), p. 55.

170

VI. AGRICULTURE

1. Paul C. Manglesdorf, Richard S. MacNeish, and Walton C. Galinat, "Domestication of Corn," *Prehistoric Agriculture*, Stuart Struever, ed. (Garden City, N.Y., 1971), p. 472.

2. Paul C. Manglesdorf, Richard S. MacNeish, and Gordon R. Willey, "Origins of Agriculture in Middle America," *Prehistoric Agriculture*, Stuart Struever, ed. (Garden City, N.Y., 1971), p. 515.

3. Robert M. Adams, "Developmental Stages in Ancient Mesopotamia," *Prehistoric Agriculture*, Stuart Struever, ed. (Garden City, N.Y., 1971), p. 579.

4. H. M. Wormington, *Prehistoric Indians of the Southwest*, 3rd Ed., (Denver, 1956), p. 37.

5. Kent V. Flannery, "Archeological Systems Theory and Early Mesoamerica," *Anthropological Archeology in the Americas* (Washington, D.C., 1968), p. 80.

6. Wormington, *Prehistoric Indians*, p. 55.

7. Alfred Whiting, *Ethnobotany of the Hopi* (Flagstaff, 1966), p. 8.

8. Edward Castetter, Willis H. Bell, and M. E. Opler, "Ethnobiological Studies in the American Southwest III: The Ethnobiology of the Chiricahua and Mescalero Apache," *University of New Mexico Bulletin*, Vol. 4, No. 5, Biological Series, 1936, p. 29.

9. Winfred Buskirk, *Western Apache Subsistence Economy*, unpublished doctoral dissertation for the University of New Mexico, 1949, pp. 115–19.

10. Wilfred W. Robbins, John Peabody Harrington, and Barbara Freire-Marreco, *Ethnobotany of the Tewa Indians*, Smithsonian Institution Bureau of American Ethnology Bulletin 55 (Washington, D.C., 1916), p. 85.

11. Odd S. Halseth, "Prehistoric Irrigation in the Salt River Valley," *University of New Mexico Bulletin*, Vol. 1, No. 5 (October 1936), p. 42.

12. Whiting, *Ethnobotany of the Hopi*, p. 11.

Beans

1. G. F. Freeman, *Southwestern Beans and Teparies*, Rev. Ed., University of Arizona Agricultural Experiment Station Bulletin 68 (January 1918), p. 40.

2. Edward Castetter and Ruth Underhill, "Ethnobiological Studies in the American Southwest II: Ethnobiology of the Papago Indians," *University of New Mexico Bulletin*, Vol. 4, No. 3, Biological Series, 1935, p. 32.

3. Freeman, *Southwestern Beans and Teparies*, p. 11.

4. *Ibid.*, p. 38.

5. *Ibid.*, pp. 28–36.

6. Vorsila L. Bohrer, "Zuni Agriculture," *El Palacio*, Vol. 67, No. 6, (December 1960), p. 189.

7. Edward Castetter and Willis H. Bell, *Yuman Indian Agriculture* (Albuquerque, 1951), p. 108.

8. Bohrer, "Zuni Agriculture," p. 190.

9. Juanita Keithly Scott, "Pinto Beans and Corn," *Mother Earth News*, Vol. 1, No. 13 (January 1972), p. 57.

10. Bohrer, "Zuni Agriculture," p. 190.

11. Wilfred W. Robbins, John Peabody Harrington, and Barbara Freire-Marreco, *Ethnobotany of the Tewa Indians*, Smithsonian Institution Bureau of American Ethnology Bulletin 55 (Washington, D.C., 1916), p. 100.

12. L. S. M. Curtin, *Healing Herbs of the Upper Rio Grande* (Los Angeles, 1965), p. 89.

Chili

1. L. S. M. Curtin, *Healing Herbs of the Upper Rio Grande* (Los Angeles, 1965), p. 64.

Corn

1. Matilda Cox Stevenson, "Ethnobotany of the Zuni," *30th Annual Report of the Bureau of American Ethnology* (Washington, 1009), pp. 30–31.

2. *Ibid.*, p. 99.

3. Wilfred W. Robbins, John Peabody Harrington, and Barbara Freire-Marreco, *Ethnobotany of the Tewa Indians*, Smithsonian Institution Bureau of American Ethnology Bulletin 55 (Washington, D.C., 1916), p. 81.

4. Alfred Whiting, *Ethnobotany of the Hopi* (Flagstaff, 1966), p. 68.

5. Edward Castetter and Ruth Underhill, "Ethnobiological Studies in the American Southwest II: Ethnobiology of the Papago Indians," *New Mexico Bulletin*, Vol. 4, No. 3, Biological Series, 1935, p. 35.

6. Leslie Spier, *Yuman Tribes of the Gila River* (Chicago, 1933), pp. 61–62.

7. Robbins, Harrington, and Freire-Marreco, *Ethnobotany of the Tewa Indians*, p. 83.

8. Frank Cushing, *Zuni Bread Stuffs*, Indian Notes and Monographs Vol. VIII (New York, 1920), pp. 186–87.

9. *Ibid.*, pp. 196–97.

10. Robbins, Harrington, and Freire-Marreco, *Ethnobotany of the Tewa Indians*, p. 83.

11. *Ibid.*, p. 342.

12. Cushing, *Zuni Bread Stuffs*, pp. 382–90.

13. *Ibid.*, p. 342.

14. Winfred Buskirk, *Western Apache Subsistence Economy*, unpublished doctoral dissertation for the University of New Mexico, 1949, p. 141.

15. *Ibid.*, p. 139.

16. Flora L. Bailey, "Navajo Foods and Cooking Methods," *American Anthropologist*, Vol. 42, No. 2 (April-June 1940), p. 281.

17. Buskirk, *Western Apache Subsistence Economy*, p. 139.

18. Miguel Hambriento, "The Foods of Old Mesilla," *New Mexico*, Vol. 25, No. 1 (January 1947), p. 49.

19. Castetter and Underhill, "Ethnobiology of the Papago Indians," p. 35.

20. Stevenson, "Ethnobotany of the Zuni," p. 100.

21. Robbins, Harrington, and Freire-Marreco, *Ethnobotany of the Tewa Indians*, p. 87.

22. *Ibid.*, p. 96.

23. *Ibid.*, p. 97.

24. William Darby and others, "A Study of the Dietary Background and Nutriture of the Navajo Indian," *The Journal of Nutrition*, Vol. 60, Supplement 2 (November 1956), p. 32.

25. Buskirk, *Western Apache Subsistence Economy*, p. 133.

26. Stevenson, "Ethnobotany of the Zuni," p. 62.

27. Robbins, Harrington, and Friere-Marreco, *Ethnobotany of the Tewa Indians*, p. 67.

Melon

1. Alfred Whiting, *Ethnobotany of the Hopi* (Flagstaff, 1966), p. 93.

2. Edward Castetter and Willis H. Bell, *Yuman Indian Agriculture* (Albuquerque, 1951), p. 129.

3. Wilfred W. Robbins, John Peabody Harrington, and Barbara Freire-
Marreco, *Ethnobotany of the Tewa Indians*, Smithsonian Institution Bureau of
American Ethnology Bulletin 55 (Washington, D.C., 1916), p. 112.

4. Leslie Spier, *Yuman Tribes of the Gila River* (Chicago, 1933), p. 65.

5. Castetter and Bell, *Yuman Indian Agriculture*, p. 128.

6. Robbins, Harrington, and Freire-Marreco, *Ethnobotany of the Tewa Indians*, p. 111.

7. Castetter and Bell, *Yuman Indian Agriculture*, p. 128.

Pumpkin

1. Grenville Goodwin, "Social Diversions and Economic Life of the Western Apaches," *American Anthropologist*, Vol. 37, No. 1 (January 1935), p. 63.

2. Winfred Buskirk, *Western Apache Subsistence Economy*, unpublished doctoral dissertation for the University of New Mexico, 1949, pp. 155–56.

3. Edward Castetter and Willis H. Bell, *Yuman Indian Agriculture*, (Albuquerque, 1951), p. 114.

4. Vorsila L. Bohrer, "Zuni Agriculture," *El Palacio*, Vol. 67, No. 6, (December 1960), p. 197.

Squash

1. Matilda Cox Stevenson, "Ethnobotany of the Zuni," *30th Annual Report of the Bureau of American Ethnology* (Washington, 1909), p. 88.

Wheat

1. Wilfred W. Robbins, John Peabody Harrington, and Barbara Freire-Marreco, *Ethnobotany of the Tewa Indians*, Smithsonian Institution Bureau of American Ethnology Bulletin 55 (Washington, D.C., 1916), pp. 107–09.

2. Tillie Burch, "In the Pueblo Kitchen," *New Mexico*, Vol. 22, No. 3 (March 1944), p. 18.

3. Edward Castetter and Willis H. Bell, *Yuman Indian Agriculture* (Albuquerque, 1951), p. 126.

4. L. S. M. Curtin, *By the Prophet of the Earth* (Santa Fe, 1949), p. 74.

5. Robbins, Harrington, and Freire-Marreco, *Ethnobotany of the Tewa Indians*, p. 109.

6. Curtin, *By the Prophet of the Earth*, p. 73.

Sources

FOODS

Santa Cruz Chili and Spice Co.
Box 177
Tumacacori, Arizona 85640

The above company carries red chili paste, dry chili powder, and canned green chilis. They are happy to have mail orders. Write for current price list.

Theo. Roybal Store
Rear 212-214-216 Galisteo Street
Santa Fe, New Mexico 87501

This old-time herb store carries an impressive line of herbs and native ingredients including panocha flour, piloncillo, blue corn meal, and pinyon nuts. Write for a complete list.

INSTRUCTIONS FOR BUILDING AN ADOBE OVEN

Stedman, Myrtle. "How to Make an Horno." *New Mexico Magazine*, August 1969, p. 33.

"Your Own Pueblo Oven." *Sunset Magazine*, August 1971, pp. 50–54.
"Pueblo Oven . . . more building, more cooking ideas." *Sunset Magazine*, July 1972, pp. 118–21.
Boudreau, Eugene H. *Making the Adobe Brick*. Berkeley: Fifth Street Press, 1971.

174

Bibliography

Adams, Robert M. "Development Stages in Ancient Mesopotamia." *Prehistoric Agriculture*. Ed. by Stuart Struever. Garden City, N.Y.: Natural History Press, 1971, pp. 572–614.

Angier, Bradford. *Wilderness Cookery*. Harrisburg, Penn.: Stackpole Books, 1961.

Baeget, Jacob. "The Aboriginal Inhabitants of the California Penninsula." *Annual Report of the Board of Regents of the Smithsonian Institution 1863.* Washington, D.C.

Bailey, Flora L. "Navajo Foods and Cooking Methods." *American Anthropologist*, Vol. 42, No. 2 (April–June 1940), pp. 270–90.

Bartlett, Katharine. "Edible Wild Plants of Northern Arizona." *Plateau*, Vol. 16, No. 1 (July 1943), pp. 11–17.

———. "Prehistoric Pueblo Foods." *Museum Notes*, Vol. 4, No. 4 (October 1931), pp. 1–4.

Basso, Keith. *The Cibecue Apache*. New York: Holt, Rinehart, and Winston, 1970.

Bell, Willis H., and Edward F. Castetter. "Ethnobiological Studies in the American Southwest VII: The Utilization of Yucca, Sotol and Beargrass by the Aborigines of the American Southwest." *University of New Mexico Bulletin*, Vol. 5, No. 5, Biological Series (December 1941).

Bergland, Berndt, and Clare E. Bolsby. *The Edible Wild*. New York: Pagurian Press Ltd., 1971.

Bohrer, Vorsila L., "Zuni Agriculture." *El Palacio*, Vol. 67, No. 6 (December 1960), pp. 181–202.

Bravery, H. E. *Successful Wine Making At Home*. New York: Arc Books, 1961.

Burch, Tillie. "In The Pueblo Kitchen." *New Mexico*, Vol. 22, No. 3 (March 1944), pp. 18–34.

175

Buskirk, Winfred. *Western Apache Subsistence Economy.* Unpublished doctoral dissertation for the University of New Mexico, 1949.

Castetter, Edward F. "Ethnobiological Studies in the American Southwest I: Uncultivated Native Plants Used As Sources of Food." *University of New Mexico Bulletin,* Vol. 4, No. 1, Biological Series (May 1935).

Castetter, Edward F., and Willis H. Bell. "Ethnobiological Studies in the American Southwest IV: The Aboriginal Utilization of the Tall Cacti in the American Southwest." *University of New Mexico Bulletin,* Vol. 5, No. 1, Biological Series (June 1937).

———. *Pima and Papago Indian Agriculture.* Albuquerque: University of New Mexico Press, 1942.

———. *Yuman Indian Agriculture.* Albuquerque: University of New Mexico Press, 1951.

Castetter, Edward, Willis H. Bell, and Alvin R. Grove. "Ethnobiological Studies in the American Southwest VI: The Early Utilization and Distribution of Agave in the American Southwest." *University of New Mexico Bulletin,* Vol. 5, No. 4 (December 1938).

Castetter, Edward, Willis H. Bell, and M. E. Opler. "Ethnobiological Studies in the American Southwest III: The Ethnobiology of the Chiricahua and Mescalero Apache." *University of New Mexico Bulletin,* Vol. 4, No. 5 (November 1936).

Castetter, Edward, and Ruth Underhill. "Ethnobiological Studies in the American Southwest II: The Ethnobiology of the Papago Indians." *University of New Mexico Bulletin,* Vol. 4, No. 3 (October 1935).

Churchill, James E. "Food Without Farming." *Mother Earth News,* Vol. 1, No. 3 (May 1970), pp. 59–63.

Curtin, L. S. M. *By the Prophet of the Earth.* Santa Fe: San Vicente Foundation Inc., 1949.

———. *Healing Herbs of the Upper Rio Grande.* Los Angeles: Southwest Museum, 1965.

Cushing, Frank Hamilton. *Zuni Breadstuff.* Indian Notes and Monographs, Vol. VIII. New York: Museum of the American Indian, 1920.

Darby, William, *et al.* "A Study of the Dietary Background and Nutriture of the Navajo Indian." *The Journal of Nutrition,* Vol. 60, Supplement 2 (November 1956).

Dodge, Natt. *Desert Wild Flowers.* Globe, Arizona: Southwestern Monuments Association, 1963.

———. *Flowers of the Southwest Deserts.* 5th Ed. Globe, Arizona: Southwestern Monuments Association, 1961.

Felger, Richard S., and Mary B. Moser. "Seri Use of Mesquite." *The Kiva,* Vol. 37, No. 1 (Fall 1971), pp. 53–60.

Flannery, Kent V. "Archeological Systems Theory and Early Mesoamerica." *Anthropological Archeology in the Americas.* Washington, D.C.: Anthropological Society of Washington, 1968, pp. 67–85.

Freeman, G. F. *Southwestern Beans and Teparies,* Rev. Ed., University of Arizona Agricultural Experiment Station Bulletin 68, Tucson (January 1918).

Gentry, H. S. "The Natural History of Jojoba (*Simmondsia chinensis*) and Its Cultural Aspects." *Economic Botany,* Vol. 12, No. 3 (July–September 1958), pp. 261–95.

Gibbons, Euell. *Stalking the Wild Asparagus.* Field Guide Ed., New York: David McKay, 1970.

Grenville, Goodwin. "Social Diversions and Economic Life of Western Apaches." *American Anthropologist,* Vol. 37, No. 1 (January 1935), pp. 55–64.

Halseth, Odd. "Prehistoric Irrigation in the Salt River Valley." *University of*
    New Mexico Bulletin, Vol. 1, No. 5 (October 1936), pp. 42–47.

Hambriento, Miguel. "The Foods of Old Mesilla." *New Mexico*, Vol. 25, No. 1
    (January 1947), pp. 16–49.

Harris, Ben Charles. *Eat the Weeds*. Barre, Massachusetts: Barre, 1971.

Hart, Elisabeth. *Pima Cookery*, 2nd Ed. Sacaton, Arizona: Cooperative Extension
    Service, 1968.

Hinton, Thomas. "A Description of the Contemporary Use of an Aboriginal
    Sonoran Food." *The Kiva*, Vol. 21, Nos. 3 and 4 (May 1956), pp. 27–28.

Kearney, Thomas H., and Robert H. Peebles. *Arizona Flora*. Berkeley: University
    of California Press, 1969.

Kirk, Donald. *Wild Edible Plants of the Western United States*. Healdsburg, Cali-
    fornia: Naturegraph Publishers, 1970.

Knochmal, A., S. Paur, and P. Duisberg. "Useful Native Plants in the American
    Southwest Deserts." *Economic Botany*, Vol. 8, No. 1 (January–March 1954),
    pp. 3–20.

Lighthall, J. I. *The Indian Household Medicine Guide*, 2nd Ed. Mokelumne Hill,
    Calif.: Health Research, 1966.

Manglesdorf, Paul C., Richard S. MacNeish, and Walton C. Galinat. "Domestica-
    tion of Corn," *Prehistoric Agriculture*. Ed. Stuart Struever. Garden City,
    N.Y.: Natural History Press, 1971, pp. 472–86.

Manglesdorf, Paul C., Richard S. MacNeish, and Gordon R. Willey. "Origins of
    Agriculture in Middle America." *Prehistoric Agriculture*. Ed. Stuart
    Struever. Garden City, N.Y.: Natural History Press, 1971, pp. 488–515.

Morton, Julia F. "Principal Wild Plant Foods of the United States excluding
    Alaska and Hawaii." *Economic Botany*, Vol. 17, No. 4 (October–December
    1963), pp. 319–30.

Nequatewa, Edmund. "Hopi Courtship and Marriage." *Museum Notes*, Vol. 5,
    No. 9 (March 1933) pp. 47–55.

———. "Some Hopi Recipes for the Preparation of Wild Plant Foods." *Plateau*,
    Vol. 16, No. 1 (July 1943), pp. 18–21.

Olsen, Larry Dean. *Outdoor Survival Skills*. Provo: Extension Publications,
    Brigham Young University, 1973.

Patraw, Pauline. *Flowers of the Southwest Mesas*. 4th Ed., Globe, Arizona: South-
    western Monuments Association, 1964.

Pfefferkorn, Ignaz. *Sonora, a Description of the Province*. Translated and an-
    notated by Theodore E. Treutlein. Albuquerque: University of New
    Mexico Press, 1949.

Robbins, Wilfred W., John Peabody Harrington, and Barbara Friere-Marreco.
    Ethnobotany of the Tewa Indians. Smithsonian Institution Bureau of
    American Ethnology Bulletin 55. Washington, D.C.: Government Print-
    ing Office, 1916.

Rombauer, Irma S., and Marion Rombauer Becker. *The Joy of Cooking*. Rev.
    and enlarged. Indianapolis: Bobbs-Merrill Co., 1964.

Ross, Winifred. *The Present Day Dietary Habits of the Papago Indians*. Unpub-
    lished master's thesis for the University of Arizona, 1941.

Russell, Frank. "The Pima Indians." *26th Annual Report of the Bureau of Amer-
    ican Ethnology*. Washington, D.C.: Government Printing Office, 1904.

Scott, Juanita Keithly. "Pinto Beans and Corn." *Mother Earth News*, Vol. 1, No.
    13 (January 1972), pp. 57–61.

Spier, Leslie. *Yuman Tribes of the Gila River*. Chicago: University of Chicago
    Press, 1933.

Stevenson, Matilda Cox. "Ethnobotany of the Zuni." *30th Annual Report of the*

BIBLIOGRAPHY

Bureau of American Ethnology. Washington, D.C.: Government Printing Office, 1908–09.

Sweet, Muriel. *Common Edible and Useful Plants of the West*. Healdsburg, California: Naturegraph Publishers, 1962.

Watson, James B. "How the Hopi Classify Their Foods." *Plateau*, Vol. 15, No. 4 (April 1943), pp. 49–51.

Watt, Mrs. J. H. "Chokecherry Wine." *Cooking in Wyoming, 1890–1965*. Cheyenne: State of Wyoming, 1965.

Whiting, Alfred. *Ethnobotany of the Hopi*. Flagstaff: Northland Press, 1966.

Wormington, H. M. *Prehistoric Indians of the Southwest*. 3rd Ed., Denver: The Denver Museum of Natural History, 1956.

Medical Index

General Index

183

190